BIOTECHNOLOGICAL POTENTIAL OF ASCIDIANS

Wacky and Weird Sea Squirts

Cover photograph credits:

Left: Didemnum mole. Photo by Silke Baron. https://creativecommons.org/licenses/by/2.0/

Top right: Sea Squirts at Bunaken Marine Park, North Sulawesi. Photo by Rob. https://creativecommons.org/licenses/by/2.0/

Bottom right: Spiky sea squirts at Lorry Bay south of Gordon's Bay on the East side of False Bay. Photo by Peter Southwood. https://creativecommons.org/licenses/by-sa/3.0/

Biology and Ecology of Marine Life Series

BIOTECHNOLOGICAL POTENTIAL OF ASCIDIANS

Wacky and Weird Sea Squirts

by

Ramasamy Santhanam, PhD

AAP | APPLE ACADEMIC PRESS

First edition published 2026

Apple Academic Press Inc.
1265 Goldenrod Circle, NE,
Palm Bay, FL 32905 USA

760 Laurentian Drive, Unit 19,
Burlington, ON L7N 0A4, Canada

CRC Press
2385 NW Executive Center Drive,
Suite 320, Boca Raton FL 33431

4 Park Square, Milton Park,
Abingdon, Oxon, OX14 4RN, UK

Library and Archives Canada Cataloguing in Publication

Title: Biotechnological potential of ascidians : wacky and weird sea squirts / Ramasamy Santhanam, PhD.
Names: Santhanam, Ramasamy, 1946- author
Series: Santhanam, Ramasamy, 1946- Biology and ecology of marine life book series.
Description: First edition. | Includes bibliographical references and index.
Identifiers: Canadiana (print) 20250143275 | Canadiana (ebook) 20250143313 | ISBN 9781774919880 (hardcover) | ISBN 9781003637127 (ebook)
Subjects: LCSH: Sea squirts.
Classification: LCC QL613 .S26 2025 | DDC 596/.2—dc23

Library of Congress Cataloging-in-Publication Data

CIP data on file with US Library of Congress

ISBN: 978-1-77491-988-0 (hbk)
ISBN: 978-1-00363-712-7 (ebk)

BIOLOGY AND ECOLOGY OF MARINE LIFE BOOK SERIES

SERIES AUTHOR

Ramasamy Santhanam, PhD
Former Dean, Fisheries College and Research Institute,
Tamil Nadu Veterinary and Animal Sciences University,
Thoothukkudi, India

Books in the Series:

- Biology and Culture of Portunid Crabs of the World Seas
- Biology and Ecology of Edible Marine Bivalve Molluscs
- Biology and Ecology of Edible Marine Gastropod Molluscs
- Biology and Ecology of Venomous Marine Snails
- Biology and Ecology of Venomous Stingrays
- Biology and Ecology of Toxic Pufferfish
- Biology and Ecology of Bioluminescent Marine Fishes
- Biology and Ecology of the Venomous Catfishes
- Biotechnological Potential of Ascidians: Wacky and Weird Sea Squirts

ABOUT THE AUTHOR

Ramasamy Santhanam, PhD

Former Dean, Fisheries College and Research Institute (FC &RI), Tamil Nadu Veterinary and Animal Sciences University (presently under Tamil Nadu Dr. J. Jayalalithaa Fisheries University), Thoothukudi, India

Ramasamy Santhanam, PhD, is the former Dean of the Fisheries College and Research Institute, Tamil Nadu Veterinary and Animal Sciences University, Thoothukudi, India. He has about 50 years of teaching and research experience in marine/ fisheries sciences. He has been serving as a fisheries expert for various government and nongovernment organizations in India and abroad. To his credit, Dr. Santhanam has 45 books on various aspects of marine life. He was a member of the American Fisheries Society, United States; World Aquaculture Society, United States; Global Fisheries Ecosystem Management Network (GFEMN), United States; and the IUCN's Commission on Ecosystem Management, Switzerland.

CONTENTS

ABBREVIATIONS

ASW	artificial seawater
BM	blood meal
BSMO	bottled sea squirt meat in oil
CAs	carbon alloys
CMC	carboxymethyl cellulose
CNCs	cellulose nanocrystals
CNF	cellulose nanofiber
CS	chondroitin sulphate
DHA	docosahexaenoic acid
EPA	eicosapentaenoic acid
FCR	feed conversion ratios
HPLC	high-performance liquid chromatography
HS	heparan sulfate
IMTA	integrated multitrophic aquaculture
MUFA	monounsaturated fatty acids
OER	oxygen evolution reaction
ORR	oxygen reduction reaction
PCs	phytochelatins
PSP	paralytic shellfish toxins
SAFA	saturated fatty acids
SGR	specific growth rate
WBR	whole-body regeneration

FOREWORD

I am immensely happy to write my Foreword for the title "Biotechnological Potential of Ascidians: Wacky and Weird Sea Squirts", a hither-to untouched aspect of Fisheries discipline written by my teacher Prof. Dr. Ramasamy Santhanam.

In recent years, the world food crisis is becoming more serious due to natural disasters such as droughts and floods, and the outbreak of the corona virus (2019- nCoV). Researchers have therefore been searching for alternative protein sources to address this global issue. Marine bio resources, such as fish and shrimp account for a large part of foods for human beings; and it is considered to be critical to tackle the global food crisis by producing cheap and high-quality protein sources. In this context, the ascidians (Subphylum,Tunicata) a group of marine animals, have been attracted a lot of attention in medical application, food and aqua feed market and water pollution issues. Several species of ascidians including the sea pineapple *(Halocynthia roretzi)* present a high nutritional value owing to their high protein levels and low calories ; and rich n-3 polyunsaturated fatty acids, namely, essential fatty acids such as EPA and DHA. They are therefore considered as a healthy seafood choice. In countries like Japan, South Korea, and Chile, wild-harvested or cultured ascidians for human consumption is on the increase. Other applications of ascidians such as fishing bait, health supplement tablets and as ornamental species for marine aquaria are also receiving greater attention.

In my opinion, this publication could serve as a potential resource for teachers and students of Fisheries Science and Marine Biology besides serving as a valuable reference for researchers on ascidians.

I congratulate the author for his timely contribution.

—Prof. S. Felix, M.F.Sc., PhD
Dean

PREFACE

Tunicates (Phylum, Chordata; Subphylum, Tunicata) include three classes of animals, namely, the benthic and sessile ascidians (commonly known as sea-squirts); and the pelagic thaliaceans, and larvaceans (or appendicularians). It is worth mentioning here that these ascidians are the only group producing pure and high-quality cellulose with excellent mechanical properties and thermal stability. Among the tunicates, the ascidians which comprise 3000 living species are the most neglected group and they have attracted the attention of researchers in recent years owing to their biotechnological applications. These ascidians are known to possess a high nutritional value and due to their high protein and low calories they are considered to be a healthy seafood choice. Further, the nutritional profile of the ascidian fatty acids is more or less similar to that of fish oil, and this makes the ascidian biomass as an ideal alternative to fish oil and fish meal in the formulation of aqua and livestock feeds. The wide varieties of biologically active secondary metabolites from cyclic peptides to aromatic alkaloids produced by different species of ascidians are considered to be potential candidates for the discovery of new drugs for treating diseases including cancer.

The nanocellulose derived from the cellulose of ascidians has been reported to be of great use in developing luminescent fibers, food packaging materials, renewable sensors, and strong composites, and these materials show significant industrial applications in material and food-related fields. Further, the polysaccharides present in the cellulose of these ascidians could be converted into glucose and further fermentations could lead to the production of bioethanol. Ascidians are also used nowadays in other applications such as in the production of health supplement tablets, fishing baits, and as ornamental species for marine aquarium.

Though several books are presently available on marine life and its biology and ecology, a new title relating to the biotechnological potential of ascidians is the need of the hour. The present title *Biotechnological Potential of Ascidians: Wacky and Weird Sea Squirts* would be the first of its kind to answer this long felt need. As blue biotechnology is gaining momentum in the recent decades, it is strongly hoped that the present title will amply provide the state of the art on the biotechnological potential of ascidians of different seas and oceans.

This new title deals with aspects such as species diversity, ecology and biology; nutritional and aqua and livestock feed values; chemical compounds and their therapeutic values; industrially important products and by-products from ascidians; ascidian cellulose as a source of second-generation biofuel production; ascidians as bioremediating agents of marine pollution; and mariculture of industrially important ascidians. It is strongly felt that this title will be of great value as a text-cum-reference for students, researchers, and teachers, of various disciplines including Marine Biology and Fisheries Science; a valuable source of reference for college and university libraries; and as a guide for the industries involved in the development of food, feed, and drugs.

I am grateful to Prof. Dr. A. Srinivasan, formerly Registrar, Tamil Nadu Dr. J. Jayalalithaa Fisheries University, India, for his productive suggestions on the text. I profoundly thank all the international scientists who gracefully shared their images for this title. The secretarial and photographical assistance provided by Mrs. Albin Panimalar Ramesh are also sincerely acknowledged.

The Author

CHAPTER 1

INTRODUCTION

Ascidians (sea squirts and sea tulips) (phylum Chordata, sub-phylum Tunicata (=Urochordata), class Ascidiacea) are exclusively marine, sessile benthic organisms of diversified sizes and shapes and are seen all over the world. A few species of sea squirts in the genus *Pyura* are commonly called sea tulips. The sea squirts get their nickname due to their ability to squirt water when they are dislocated from their habitat, namely, sea. These ancestral chordates (or invertebrate chordates) are placed in a very unique position between invertebrates and vertebrates as they contain the vertebrate feature, namely, a notochord in their swimming larval stages; and as adults, they are invertebrate filter feeders and are immotile. The filter-feeding habit of adult ascidians allows them to accumulate pollutants. So by observing ascidian populations, one can get a good indication of pollution levels of the environment (https://australian.museum/learn/animals/sea-squirts/).

These ascidians (=sea pork) are abundant in harbors. In these areas, they are known to occur in protected habitats such as kelp forests and coral reefs; and between sandy and rocky shore areas (Watters, 2018).

Certain species of ascidians have also been reported from euryhaline brackish-water marine ecosystems including mangroves. A total of 3020 extant species of ascidians have so far been described (https://blogs.ifas.ufl.edu/brevardco/2014/05/12/tunicates-aka-sea-squirts/) and they are solitary or colonial and are in many colors, shapes and sizes, and colors. Among the species of ascidians, more than 60% are colonial (Wawrzyniak et al., 2021a; Shenkar and Swalla, 2011). The ascidians are generally grouped into three orders based on the structure of their branchial sac. They are exclusively colonial forms (Aplousobranchia); and both solitary and colonial forms (Phlebobranchia and Stolidobranchia). The adult branchial sacs of these three branchial sacs are simple, vascular, and folded, respectively. All ascidians

Biotechnological Potential of Ascidians: Wacky and Weird Sea Squirts. Ramasamy Santhanam (Author)

are filter feeders using their two siphons, namely, inhalant (incurrent) and exhalent (excurrent) siphons. While the inhalant siphon takes in water, when food and other particles are extracted, the exhalent siphon expels water out. Tunicates earn their name from their somewhat flexible and tough outer covering, that is, organic exoskeleton called "tunic" which forms 45% of total animal dry weight (as in the case of *Ciona intestinalis*). This tunic is composed of several layers of cellulose which is made of a polysaccharide with several sugar molecules, protein, and ash, and this tunic acts as a skeletal structure. On the other hand, the inner animal which forms 55% of total animal weight consists mainly of protein and ω-3 fatty acids (omega-3 fatty acids) and this part of the animal may be used as feed ingredients. It is also worth mentioning here that the ascidians are the only animal source of cellulose (Seddiqi et al., 2021). Many tropical species of ascidians have been reported to harbor single-celled microalgae particularly cyanobacteria (blue-green algae) as symbionts inside their tissues. It is interesting to note that *Ascidia nigra* and *Ecteinascidia turbinata* possess other organisms such as amphipods, copepods, crabs, and brittle-stars on or inside their tunic. The ascidians are hermaphrodite (both asexual and sexual reproduction) organisms. Among the chordates, the colonial ascidians are the only group with the characteristic feature of whole-body regeneration (WBR). Using this WBR, a minute fragment of ascidian vasculature could be made into a fully functional adult in a period of about 10 days. Further, these ascidians have been reported to be unique in reversing their blood circulation. Using this technique, they can change the direction of the blood flow by reverse peristalsis at regular intervals (https://www.toppr.com/ask/question/ascidian-heart-is-unique-amongst-chordates-in-sense-that/). Further, the "ascidian heart" does not function before metamorphosis. Though there are reports about the nil participation of these organisms in marine food chain due to their mantle, fish, flatworms, sea urchins, and sea stars are believed to be the sea squirts' primary predators.

1.1 BIOTECHNOLOGICAL APPLICATIONS OF ASCIDIANS

Owing to their unique biomaterial contents such as protein–polysaccharide complexes (e.g., chitin, scleroprotein, iodine-binding proteins, elastin, collagen, and glycosaminoglycan), amino-sugars, and cellulose, these ascidians have been reported to be of great importance in cellulose nanomaterial production, medical application, food market, and water pollution issues (as bioremediating agents).

Human food: Although the ascidians are rarely eaten by humans, they form an important link in the food chain and thus they indirectly provide humans with a source of food (https://www.britannica.com/animal/tunicate). Edible ascidians such as *C. intestinalis*, *Halocynthia aurantium*, and *Halocynthia roretzi* present a high nutritional value with low calories and high protein levels. They are also a good source of n-3 polyunsaturated fatty acids (PUFAs) such as eicosapentaenoic acid (EPA) and docosahexaenoic acid (DHA), and the composition profile of these fatty acids is more or less similar to that of fish oil. Therefore, these ascidians are considered to be a healthy seafood choice for humans (Marques et al., 2022). Watters (2018) reported that these ascidians are farmed and consumed in several countries such as Korea, Chile, Japan, and France (Watters, 2018).

Eating ascidians reverses signs of cognitive decline and aging: Supplementing a diet with ascidians has been reported to reverse some of the main signs of aging in animal models. Ascidians are known to contain glycerophospholipids, namely, plasmalogens, which are essential to human body processes and are present all over human bodies naturally, particularly in the heart, brain, and immune cells. As one gets older, the amount of this substance decreases in human body causing many neurode- generative diseases, namely, Alzheimer's disease, and Parkinson's disease. Supplementing plasmalogens derived from ascidians in the diet of aged mice yielded profound effects on the physical symptoms and learning abilities of these experimental mice. Research findings state that ascidian-derived plasmalogens may not only stop cognitive decline, but may also reverse cognitive impairments in the aging brain. Further, aged mice fed with the plasmalogens were found to grow new black hair which was thicker and glossier than aged mice not fed with plasmalogens (https://neurosciencenews.com/aging-sea-squirts-20570/, 2022).

Aquafeed: The ascidians biomass may also be a potential alternative to commercial fish oil and fish meal in commercial fish and shrimp feeds usable in mariculture practices. Further, as the amino acid composition of ascidians has been reported to be similar to that of egg albumin, this ascidian meal can also be a feed source for ornamental fishes and other aquariculture operations, for example, partially or fully replacing of commercial fish meal with the tunic (cellulose) of the tunicate, *H. roretzi* in the aquafeeds for the mollusk, *Haliotis discus* yielded promising results (Marques et al., 2022).

Fishing bait: The edible species of ascidians, namely, *Pyura pachydermatina* and *Pyura praeputialis* are presently used as fishing bait in New Zealand and Australia, respectively (Lambert et al., 2016).

Pharmaceutical and biomedicals: The study of ascidians has attracted several researchers due to their unique commercial applications and this is largely due to their pharmaceutical and biomedical potentials of these organisms. About 1000 bioactive molecules with antiviral, antimicrobial, anti-inflammatory, and antitumoral activities have so far been derived from these ascidians (Wawrzyniak et al., 2021b). One FDA-approved drug, namely, Ecteinascidin (ET-743, trabectedin) (trade name Yondelis), and another, Australian Regulatory Agency–approved drug, namely, Aplidin, are presently marketed for the treatment of cancer (https://pharmacy.ucsf.edu/news/2021/02/cancer-drug-stops-sars-cov-2-lab; Watters, 2018). Ascidians have been reported to be capable of re-engineering their transcriptome which is largely by the process of RNA trans-splicing and alternative splicing. This rewriting RNA (RNA-editing technology) can be of great use for early diagnoses and effective treatment of cancer. Inspired by this characteristic of ascidians, the "Ascidian Therapeutics" has been floated to the surface with $50 million for therapeutic benefit (https://www.fiercebiotech.com/biotech/ascidian-therapeutics-rises-atps-ocean-floor-50-million-and-dynes-former-ceo). Further, this technology is believed to help in expanding the therapeutic potential of RNA medicines for cancer patients who are awaiting breakthroughs. Further, several species of colonial ascidians have appeared as model organisms in a number of research fields such as angiogenesis, immunobiology, allorecognition, and aging. Further, the compounds such as alkaloids, peptides, polyketides, glycosaminoglycans, sterols, collagens, and sulfated polysaccharides can be exploited from these tunicates as potential byproducts in the chemical and drug industries due to their antitumor, antibacterial, antifungal, and anti-inflammatory activities. Furthermore, about 160 molecules produced by the different species of tunicates and their symbiotic microbes have been reported to possess antimicrobial activity (Marques et al., 2022).

Biosurfactants: The bacteria associated with ascidians form a potential source of biosurfactants with applications in industries related to food processing (Marques et al., 2022).

Marine-integrated multitrophic aquaculture (MIMTA): As ascidians grow very quickly and are re-harvested two or three times a year, they may be incorporated in integrated multitrophic aquaculture (IMTA) framework. IMTA framework with ascidian (*Styela clava*)—sea cucumber (*Apostichopus japonicus*)—microalgae (cyanobacteria) has not only optimized the sea cucumber's growth but also had the ability to substantially mitigate the polluting sediment, reduce the harmful bacteria, and remove the nutrients like phosphorus and nitrogen thereby the water body is also purified considerably. It is also suggested that multiple combinations of different species of ascidian with other animals such as fish, shellfish, or echinoderms in IMTA frameworks may enhance socio-economic and environmental performance (Marques et al., 2022).

Detoxification: The unique presence of short Cys-rich peptides, namely, phytochelatins (PCs) in the tissue extracts of the ubiquitous solitary ascidian, *C. intestinalis* can play a significant role in heavy metal control and decontamination through metal chelation. As these PCs are capable of binding to lead ions leading to their sequestration, they could also play an important role as components of detoxificating mechanism (Dou and Dong, 2019).

Bioindicators of micropollution and bioremediating agents: Ascidians have also been reported to be of great use in environmental bio-surveillance. As many species of ascidians are capable of tolerating and accumulating heavy metals, such species could be used as valuable bioindicators of water quality. Further, the embryos of ascidian could also be useful as models in studies relating to neurodevelopmental toxicity of various chemical compounds (Watters, 2018).

Industrial applications: The ascidians are believed to be the only animals producing cellulose among the animal kingdom (Dou and Dong, 2019). The applications of cellulose in the industrial products are well known. Further, the solitary and colonial ascidians' larval bio-adhesion properties have helped in the development of fouling-resistant materials and man-made glues. Ascidians are also nowadays used in other industrial applications, namely, health supplement tablets, fishing bait, and as marine aquarium species (Marques et al., 2022).

Presently, the tunics of certain species of ascidians are utilized for cellulose production and their body tissues serve as human food. Intensive research is therefore needed to explore the possibilities of utilizing all ascidian species' inner body tissues as human foods, tunics as sources of animal feed, and cellulose as dietary fiber. Further, as many species of ascidians possess sulfated polysaccharides, collagens, sterols, glycosaminoglycans, and trace elements, they offer vast scope for the exploration of byproducts for pharmaceutical, cosmetic, and fine chemical industries (Zhao and Li, 2016). Further, the role of ascidians as an alternative to fish meal in the formulation of cost-effective aquafeeds is encouraging and further intensive research is needed on this aspect. Owing to their removal of heavy metals and other pollutants through filtration, the ascidians could play a significant role as bio-remediators of marine pollution. Further, many tunicate species could serve as potential species for the MIMTA frameworks to help in the production of edible and anticancer species of ascidians. Popularization of MIMTA is the need of the hour, especially in countries where they are not adequately utilized. Though only a few species of tunicates serve as commercially valuable sources of sea food in some countries, their utilization is yet to become popular like other sea food in several countries. This may possibly be due to the conservative food habits of the people. As several methods of utilization of edible ascidians are readily available presently, these marine organisms offer great potential to develop as high-value nutritional products.

CHAPTER 2

SPECIES DIVERSITY, ECOLOGY, AND BIOLOGY OF ASCIDIANS

The early ancestral forms of chordates are the protochordates (proto = first) (=prochordata). It is an informal group under the phylum Chordata, and it is not included in any of the taxonomic categories. In the earlier classifications, the prochordates included three subphyla, namely, Cephalochordata, Hemichordata, and Urochordata. However, this system of classification is not followed presently. According to WoRMS, Hemichordata is presently assigned to a separate invertebrate Phylum (https://homework.study.com/explanation/why-are-urochordata-cephalochordata-called-protochordata.html). Also the phylum Chordata presently includes subphyla Cephaochordata, Urochordata, and Vertebrata. The present taxonomy of the ascidians along with the characteristics of different classes and orders is given below.

The subphylum Tunicata (=Urochordata) of the phylum Chordata includes the class Ascidiacea which contains all the ascidians in two subclasses, namely, subclass Enterogona and subclass Pleurogona. While the former is with the order Aplousiobrancia and order Phlebobranchia, the latter has only one order, namely, Stolidobranchia. The chief characteristics of the subphylum, subclasses, and orders of the class Ascidiacea are given below.

Subphylum Tunicata (=Urochordata): Animals of this subphylum are chordates with notochord which is, however, restricted to the tail of their tadpole larvae. The body is covered with an outer covering made of tunic containing cellulose. The atrium is present and is opening dorsally. The heart is present and coelom is reduced. There are no clear traces of segmentation.

Class Ascidiacea (sea squirts): It includes all solitary or colonial ascidians which are sessile and possess a tadpole larva with the characteristic chordate

Biotechnological Potential of Ascidians: Wacky and Weird Sea Squirts. Ramasamy Santhanam (Author)

features of a notochord and a dorsal, hollow nerve cord. Oral and atrial apertures of these animals are usually directed away from the substrate.

Subclass Enterogona: In these tunicates, the gonads are without pairing. Body is divided into a thorax and an abdomen.

Order Aplousobranchia: In these ascidians, the gills are without any fold and longitudinal vessels or bars are lacking in these gills. The digestive tract and genital organs are located in the posterior region of body.

Order Phlebobranchia: In these ascidians, the gills have longitudinal vessels and bars but without folds. The gonads are located on one side, that is, near digestive tract.

Subclass Pleurogona: In these ascidians, the gonads and digestive tract are seen by the gill side.

Order Stolidobranchia: In these ascidians, the longitudinal vessels of the gills are folded. Further, the pharynx enters atrium through a pair of slit-like openings.

2.1 SPECIES DIVERSITY OF ASCIDIANS IN WORLD SEAS

The studies made on the systematic division of world ascidian species include 2,815 valid species and the order Aplousobranchia of the class Ascidiacea has the maximum number of families and species (i.e., 1480). As per the World Database, presently, there are 26 families in the class Ascidiacea, and the order Aplousobranchia has 13 families and its family Didemnidae has the highest number of species (578). Further, the genera *Aplidium* and *Didemnum* of the order Aplousobranchia have the highest number of described species, namely, 259 and 228, respectively. The number of species in the different families of the orders of the class Ascidiacea is shown in Table 2.1.

2.1.1 GLOBAL DISTRIBUTION OF ASCIDIANS

Among the different oceanic areas, the Indo-Pacific recorded the maximum number of species (723) followed by the Mediterranean (229), Atlantic, and Pacific coasts of North America (170), and Antarctica (107) (Shenkar and Swalla, 2011). The areas of rich, moderately abundant, and poor diversity of ascidians in the world seas are shown in Tables 2.2–2.4.

TABLE 2.1 Species Diversity of Ascidians

Order/Family	No. of Valid Genera	No. of Valid Species
Aplousobranchia (order) Clavelinidae (=Pycnoclavellidae)	13	48
Diazonidae	2	38
Didemnidae	8	578
Euherdmaniidae	1	13
Holozoidae	4	102
Placentelidae	1	3
Polycitoridae	7	187
Polyclinidae	8	412
Protopolyclinidae	3	7
Pseudodistomidae	3	32
Ritterellidae	1	31
Stomozoidae	1	5
Vitrumidae	1	1
Phlebobranchia (order) Agneziidae	6	32
Ascidiidae	16	138
Cionidae	3	17
Corellidae	16	41
Dimeatidae	1	2
Hypobythiidae	1	2
Octacnemidae	10	26
Perophoridae	5	50
Plurellidae	2	7
Stolidobranchia (order) Hexacrobylidae	4	13
Molgulidae	39	227
Pyuridae	34	245
Styelidae	77	535

Source: Adapted from Shenkar and Swalla (2011).

2.2 ECOLOGY OF ASCIDIANS

Ascidians are in three different forms, namely, solitary (single form), colonial (colonies of individuals), or compounds (tunicates attached together, while having common tunic). Adult ascidians are sessile and are generally inhabiting a wide variety of intertidal habitats such as soft sediments, coral reefs, rocky substrates, sea grasses, and algae in shallow waters of 0–10 m.

TABLE 2.2 Areas of Rich Ascidians in the World Seas

Area/Region	No. of Ascidian Species
Australia	717
New Caledonia	317
Tasmania	249
Mediterranean Sea	229
West Pacific (Palau, Papua New Guinea, Indonesia, and the Philippines)	187
South Africa	168
West Mediterranean	165
Japan	163
New Zealand	124
Guam	117
Antarctica	107
Indo West Pacific	102

Source: Adapted from Shenkar and Swalla (2011).

TABLE 2.3 Areas of Moderately Abundant Ascidians in the World Seas

Area/Region	No. of Ascidian Species
French Polynesia	92
India	88
South America	87
East Mediterranean	86
Adriatic	85
NW Pacific	80
Gulf of Mexico	70
Gibraltar	66
Iberia	64
Fiji	60
Pacific NW	60
Panama	58
British	58
Timor and Arafura Sea	57
Brazil	56
Africa (NW)	55
Chile	55

Source: Adapted from Shenkar and Swalla (2011).

TABLE 2.4 Areas of Poor Ascidians in the World Seas

Area/Region	No. of Ascidian Species
Scandinavia	48
Red Sea	47
Hawaii	45
California	45
Belize	40
Jamaica	39
Cuba	39
Gulf of Saint Lawrence	37
Florida	36
Hong Kong	31
Bermuda	31
West indies	31
Tanzania	31
Mozambique	29
Venezuela	29
Massachusetts	26
Gulf of Aqaba	25
Azores Islands	19
Galapagos	16
Bering Sea	16
Bahrain	15
Bay of Fundy	12
Black Sea	10

Source: Adapted from Shenkar and Swalla (2011).

Gao et al. (2023) reported that certain species of ascidians occur even at depths up to 6000 m; and they exhibit robust growth and proliferate even on boat bottoms, marine buoys, floats, and pilings with mild wave action and rich nutrients. Further, these ascidians have been reported to prefer cold weather and live mainly in cold or temperate regions; and in tropical regions fewer species at smaller sizes. These organisms are also known to foul various artificial substrata such as jetties, ship hulls, floating docks, and so on. These organisms cannot avoid salinity or temperature changes throughout their adult life, once they remain sessile following larval settlement. Though most of the ascidians prefer seawater salinities, a few species may survive in brackish-water or mangrove salinities below 20–25‰ or above 44‰ as in enclosed coastal lagoons or lakes. For example, the tropical ascidian, *Ecteinascidia thurstoni* has been reported in habitats with salinities reaching 46‰. Similarly, species such as *C. intestinalis* have been reported to survive in a wide range of salinities (12–40‰). Further, ascidians are believed to show a wider tolerance to temperature range than salinities, for example, Antarctic species of ascidians can tolerate temperature as low as −1.9°C, while other species can tolerate seawater temperatures higher than 35°C (Shenkar and Swalla, 2011).

2.3 BIOLOGY OF ASCIDIANS

Ascidians are either solitary or colonial. While the solitary ascidians live independently of their tunics, colonial ascidians live physically attached to each other's tunics. Colonies possess a few or many zooids which are either partially isolated or connected by stolons. These zooids have a wide size range. Size of solitary forms (e.g., sea vase, *C. intestinalis*) may range from 1 mm to over 20 cm in length. In the colonial, sea grapes (*Molgula manhattensis*), the zooids tend to group together. It is very amazing to note that the star tunicate, *Botryllus schlosseri* is found crowded on rocks with star-like shapes. The individual zooids of these colonial sea squirts either share an excurrent siphon or they are connected by a common tunicate (https://www.whoi.edu/science/B/people/kamaral/SeaSquirts.html). Ascidians are known to have a wide range of coloration from translucent to blue, brown, green, yellow, and red; and their life span may be from a few months to a year (Gao et al., 2023).

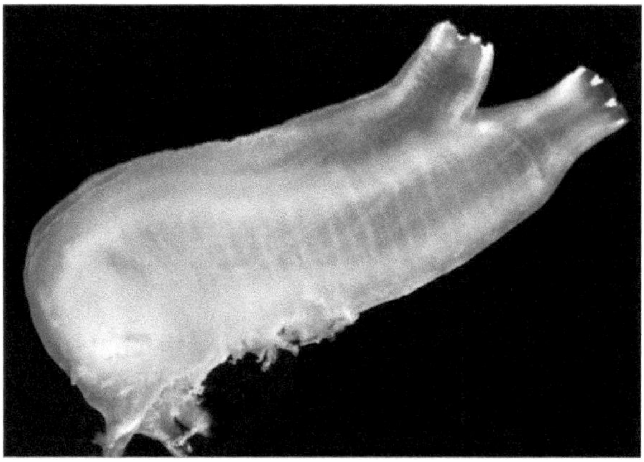

Ciona intestinalis

Image credit: E. A. Lazo-Wasem, https://creativecommons.org/publicdomain/zero/1.0/

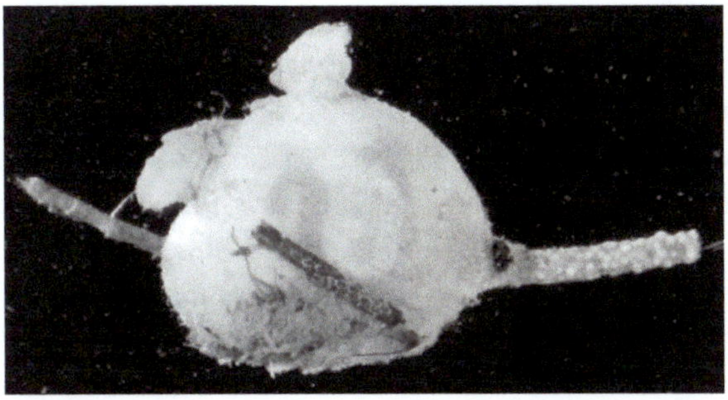

Molgula manhattensis

Image credit: Smithsonian Environmental Research Center - Molgula manhattensis_ Manokin_08-28-17_0611. https://creativecommons.org/licenses/by/2.0/

Botryllus schlosseri

Image credit: Eric A. Lazo-Wasem, https://creativecommons.org/publicdomain/zero/1.0/

Morphology: Body of the animal is bag-like. While its basal end is attached to the substratum, the free end is directed upward. There are two openings, namely, an eight-lobed oral funnel (=inhalent or incurrent siphon) at the top and a six-lobed atrial funnel (=exhalent or excurrent siphon) which is situated laterally at a distance from the free end. In the middle of the atrial siphon, there is an atriopore. Body is covered by a tunic or test which is tough but translucent. Water is known to enter the body of the animal through the oral siphon and go out through the atrial siphon. During this process, the mucus-coated pharynx not only filters the water but also collects food particles such as microalgae. Further, oxygen is also simultaneously drawn from the collected water as it passes via the gill clefts.

Body wall, atrial cavity, and coelom: In an ascidian, its outer tunic and inner mantle constitute its body wall. Tunic is secreted by epidermis and is made of the chemical, tunicin which is nothing but glycoprotein and cellulose. Though the cells of the tunic are of mesodermal in origin, they are known to migrate to the outer surface through the ectoderm. Further, the outgrowths of the body wall find their way to the test and get branched in all directions. A longitudinal septum bifurcates a channel into two and each one of them ends in a small bulb. Blood goes along one passage and returns via the second passage. During this process, the blood gets circulated in the bulb. Mantle, the soft body wall of the animal is situated under the test. This mantle is formed by the ectoderm cells with the connective tissue and their muscle fibers. Further, this mantle gets suspended in the test and is well attached to it only round the atrial and oral siphons. In several species, the mantle forms two short but wide tubular projections, namely, the oral and atrial siphons which are closed by strong sphincter muscles.

Mantle encloses the atrial or peribranchial cavity (atrium) which opens to the exterior through an atriopore. Atrial cavity is formed by an involution of the outer surface. Coelom is very much reduced and is invariably restricted to pericardial and gonadal cavities.

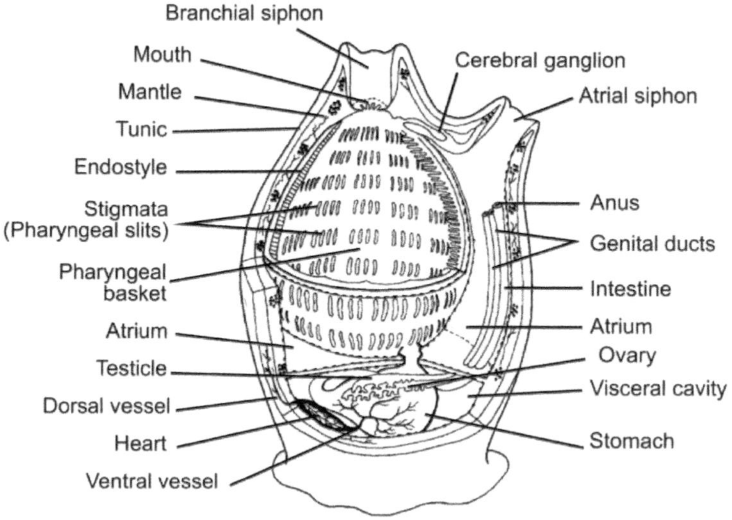

Internals of a typical ascidian.

Image credit: Jon Houseman, https://creativecommons.org/licenses/by-sa/3.0/

Properties of tunic (outer covering) of ascidians: The body of ascidians is covered with a tunic, which possesses a complex structure. On the outside, this tunic is covered with a thin but hard cuticle; under this cuticle, there is a layer of cells containing tunicin. This tunicin is a composite structure of cellulose fibrils connected to a protein matrix and are attached together through serine, threonine, and the other hydroxylated amino acids. However, in some ascidians, cellulose may be attached to either proteins or protein–polysaccharide (i.e., proteins and acid mucopolysaccharides). The ascidian cellulose has been reported to possess a crystal structure which is similar to the plant cellulose, but with a higher crystallinity. Supportive structures, such as collagen, glycosaminoglycans, chitin, elastin, and pseudoelastin may also be present in the ascidian tunic. Further, scleroprotein, iodine-binding proteins, and thyroxine have also been reported in the cellulose of some species. Under the tunic, there is a skin–muscular sac which is consisting of a single-layer epithelium and transverse and longitudinal muscle sacs fused with it (Kononova et al., 2022; Hassanzadeh, 2011).

Digestive system: In ascidian, mouth is surrounded by a velum which is located at the posterior end of oral siphon. Velum bears tentacles. Actions of the mouth are regulated by the sphincter muscles. Mouth leads to a short but wide oral passage, the stomodaeum, which communicates with a large branchial chamber or pharynx. Pharynx extends posteriorly almost nearer to the base of the body and is found attached to the mantle along the ventral side. Atrium is the space between the mantle and pharynx. Wall of the pharynx has a number of slit-like vertically disposed openings called stigmata which are arranged in transverse rows. Edges of the stigmata have a number of strong cilia. At the end of the metamorphosis, two protostigmata, that is, primary gills are located on each side of the pharynx and the number of gills gets increased to six on each side through the development of independent perforations. Due to growth, these protostigmata get elongated and subdivided to form a new row of definitive stigmata. Further, as growth proceeds, additional rows of stigmata are formed thus a perforated pharynx is produced. Inner surface of the pharynx has an endostyle which is a grooved thickening and takes the shape of a pair of ventral longitudinal folds.

On each side of the endostyle, there are two rows of mucous cells which are separated by a row of ciliated cells. Further, a group of cells with long cilia is also present in the endostyle's median groove. The dorsal lamina, a ciliated fold, is found on the inner surface of the pharynx. The pharynx serves as a food-capturing and respiratory organ, and it leads to a short and narrow esophagus which is located at the posterior end of dorsal lamina.

The mantle present on the left side of the animal occupies the esophagus, stomach, and intestine. The stomach is a large, fusiform organ and is with a thick glandular wall. On the other hand, the intestine is a narrow, thin-walled structure. A thickening, that is, typhlosole is seen along the inner wall of the intestine. Delicate ramified tubules or the pyloric glands which are seen on the wall of the intestine open into the stomach via a duct.

Feeding and predation: All the ascidians are almost filter feeders on phytoplankton and other small particles. Cilia present in the stigmata create water current through their beating movements. The animal picks up its food particles when the water enters the oral or buccal siphon, passes via the branchial basket containing gill slits, and exists through the excurrent siphon. The endostyle present in the branchial basket helps in trapping the food particles by its secreted mucus. Food particles such as microalgae (plankton) and detritus brought by the water current get entangled in this mucus sheet. The dorsal lamina present along the mid-dorsal line of the pharynx gathers the trapped food from the mucus and sends them to the esophagus. There are several deep-sea species of predatory ascidians, especially in the family Octacnemidae (Order, Phlebobranchia). In these ascidians, the enlarged oral siphon serves as the mouth which is believed to capture large prey. Adult ascidians form the food of fish and other carnivores. In order to avoid predation, these ascidians have been reported to possess a variety of chemical defenses which are believed to be synthesized by the symbiotic bacteria. These bacteria are known to deter the predators; and some species of ascidians could concentrate the toxic heavy metals including vanadium, which deter predatory fish (Holland, 2016). Gao et al. (2023) reported that as omnivorous filter feeders, these ascidians may also employ their circular muscle bands to pump the water via their oral siphons to filter water for collecting food particles. It is also reported that these ascidians trap the food particles of 1 μm–1 mm size. Further, the food trapping, fine mucous mesh of these ascidians is capable of filtering water at about 1000 times the volume of the organism with a filtration rate of 15.3 mL/s. The food of ascidians consists of 80% phytoplankton with diatoms, silicoflagellates, and coccolithophorids; and the protozoan radiolarians and tintinnids form the rest (20%) (Gao et al., 2023).

Digestion and absorption: Both the digestive and excretory functions are done by the pyloric glands of the stomach and absorption takes place finally in the intestine.

Respiration: The walls of the stigmata which are highly vascularized perform the respiratory function. These stigmata absorb the dissolved oxygen brought by the inward water current and carbon dioxide so released is sent out by the outward current.

Vascular system: The heart of the ascidian resembles a fusiform muscular sac and is located near the stomach. From the ventral part of the heart, the branchio-cardial vessel runs along the mid-ventral line of the pharynx. This vessel sends off branches which run between the rows of stigmata. A cardio-visceral vessel arising from the dorsal end of the heart sends several branches which are found distributed on the digestive organs. All such vessels open into the viscero-branchial sinus which communicates with the transverse branchial vessels. Apart from the primary vessels, several lacunae are also found extending throughout the body including the tunic. In the functioning of the heart, the peristaltic contraction takes place from one end of the heart to the other end for a known period. The next contraction which is after a pause takes place in the opposite direction. In these ways, the direction of the blood of the heart is reversed at regular intervals. The plasma of blood is colorless with few corpuscles. In the blood, lymphocytes, phagocytes, macrophages, and so on are present.

Excretory system: In ascidians, the nephrocytes of the blood are concerned with excretion. Urate and xanthine particles present in the cells are sent out in renal organs or specialized excretory vesicles in the form of concretions.

Nervous system: In ascidians, its simple nervous system consists of a neural, dorsal ganglion which is found embedded in the mantle. This ganglion, a dorso-ventrally elongated structure gives off nerves at each end thereby innervating the different parts of the body. Further, the nervous system has also a neural gland which is situated on the neural ganglion's ventral side. A folded duct from this neural gland runs forward and it opens in to the pharynx via a ciliated funnel. This duct forms a prominent dorsal tubercle and is projecting into the pharyngeal cavity.

Reproduction and development: All the ascidians are hermaphroditic (i.e., sexes are united). Ovary and testis are closely located on the left side of the body in the intestinal loop. The gonoduct (oviduct or spermiduct) which is long and continuous with the gonad opens in the atrium near the anus. Self-fertilization occurs only in a few species and it is prevented in several species and this is due to the release of the sperms and ova in different times. Ascidians release small, yolkless eggs which float on water and get fertilized externally. Development is indirect with the associated metamorphosis. Holoblastic segmentation takes place. Only a few fertilized cells yield the blastula and gastrulation is largely by the process of invagination. In about three days after fertilization, the elongated gastrula is transformed into a fish-like tadpole larva. In *Ciona*, the tadpole larva appears within 24 h, depending on the prevailing temperature (The total period from embryo to fertile adult is believed to be just three months). The tadpole larva is known to swim for a short period till they find a suitable substratum to which it attaches by its anterior adhesive papillae. Subsequently, the tail of the larva gets resorbed. Ascidians have been reported to normally reproduce sexually at regular intervals during their life span, which may be a year or more (Holland, 2016). Gao et al. (2023) reported that colonial species release their mature larvae into the water, while most of the other species spawn their eggs.

Characteristics of the tadpole larva of ascidians: The tadpole larva is highly motile due to its tail and this larva is known to starve for some time. Body of the larva is covered by tunic and it has an oval-shaped head and a long tail. Front of the head bears three adhesive (one mid-dorsal and two ventrolateral) papillae or chin warts. Tail is laterally compressed and is pointed terminally. It has a caudal fin which is continuous with dorsal and ventral fins. The central nervous system of this larva consists of a large anterior sense vesicle which is opens in the pharynx. A median eye is present in this larva. An endodermal notochord present in the tail region is found extending up to the pharynx. The tail region of the larva has segmental muscles which are arranged on both sides of the nerve cord. The alimentary canal of this tadpole larva has two regions, namely, anterior region which is consisting of a mouth, a pharynx with two pairs of gill slits and an endostyle. The posterior region of this larva occupies esophagus, stomach, and intestine. The atrial sacs of the larva are paired. This larva has been reported to be negatively geotactic and positively phototactic. This larva becomes sluggish after an active period and it starts looking for an ideal substratum for its settlement. Then, it fixes itself with its adhesive papillae and metamorphosis begins now.

Metamorphosis of the tadpole larva: The metamorphosis of tadpole larva in ascidians takes place in two forms, namely, retrogressive and progressive. During the retrogressive metamorphosis, majority of the organs of the larva are lost and in the progressive metamorphosis, some structures develop. During the retrogressive metamorphosis, the tail length is believed to be considerably reduced with subsequent disappearance of the tail. During this stage, the notochord gets coiled and disorganized to disappear with the tail. Further, the nerve cord gets restricted to the trunk region with the final appearance of a solid ganglion only. During the progressive metamorphosis, the pharynx enlarges in size with considerable increase in stigmata. Further, its atrium gets enlarged and the reproductive organs develop gradually. In nut shell, the free-swimming tadpole larva loses its notochord, nervous system and organs of special senses; and it becomes a nonmotile adult during its settlement. In its adult stage, only the gill slits, organs of feeding and endostyle are present. (Shriya, https://www.notesonzoology.com/subphylum-urochordata/ascidia-external-features-and-metamorphosis/6271).

Regeneration of body parts: The ascidians are known to regenerate all or part of their body. Regeneration of these ascidians is very much similar to their asexual reproduction by budding. Though the solitary ascidians do not bud, solitary species like Ciona, are also known to regenerate certain organs of their body. As ascidians cannot escape from their predators like fish, this regeneration is particularly useful to them. For example, if a fish bites off part of a siphon of an ascidian, this siphon can be regenerated (Holland, 2016).

Associations: Ascidians are believed to host various parasitic animals. Further, some tropical ascidians have symbiotic association with microscopic plants and cyanobacteria (blue-green algae) which that may provide them with food.

Predatory animals of ascidians: Cartilaginous fishes such as sharks and skates are known to eat sea squirts.

2.3.1 ASEXUAL REPRODUCTION IN ASCIDIANS

In the different groups of ascidians, there are several modes of asexual reproduction such as stolonic budding, vascular budding, palleal or peribranchial budding, budding by strobilation and pyloric budding (Gasparini and Ballarin, 2018).

2.3.2 ASCIDIAN SPECIES WITH ASEXUAL REPRODUCTION THROUGH BUDDING

Asexual reproduction through budding has been observed in the following species of ascidians (Berrill, 1935; Kurn et al., 2011).

Family, Diazonidae: *Diazona violacea* and *Tylobranchion speciosum*

Family, Synoicidae: *Morchellium argus*, *Aplidium elegans* (=*Sidnyum elegans*), and *Euherdmania claviformis*

Family, Distomidae: *Archidistoma aggregatum*, *Eudistoma olivacea*, *Cystodytes lobatus* (=*Eudistoma lobatum*), *Distaplia rosea*, *Distalpia magnilarva*, *Nephtheis fascicularis* (=*Colella thomsoni*), and *Distalpia concreta* (=*Colella concreta*)

Family, Didemnidae: *Diplosoma listerianum* var. *gelatinosum*

Family, Clavelinidae: *Clavelina lepadiformis*, *Clavelina picta*, and *Chondrostachys cylindrica*

Family, Perophorida: *Perophora listeri*, *Perophera viridis*, *Perophora bermudensis*, *Ecteinascidia conklini*, and *Ecteinascidia turbinata*

Family: Polyclinidae: *Aplidium pallidum*, *Amouroecium yamazii*, and *Amouroecium constellatum*

Family, Styelidae: *Eusynstyela tincta* (=*Polyandrocarpa tincta*), *Stolonica socialis*, *Distomus variolosus*, *Symplegma viride*, *Botrylloides leachi*, *Botryllus schlosseri*, *Botrylloides violaceus*, and *Polyandrocarpa misakiensis*.

2.3.3 FUNCTIONAL TYPES OF BUDDING

Two functional types of budding, namely, propagative and survival, have been reported in these ascidians. While the propagative budding is active and it functions in colony growth, the survival budding is passive and it serves primarily in the animal's survival during adverse environmental conditions (Nakauchi, 1982).

2.3.4 PROPAGATIVE BUDDING

The species of colonial ascidians possessing propagative budding include *Botryllus primigenus*, *B. schlosseri*, *B. violaceus*, *Botrylloides digense*,

Diplosoma gelatinosum, *Distaplia unigermis*, *Hypsistozoa fasmeriana*, *Leptoclinum mitsukurii*, *Perophora japonica*, *Polyzoa vesiculiphora*, *Polyandrocarpa stolonifera*, and *Polyandrocarpa vesiculiphora* (Nakauchi, 1982). Gasparini and Ballarin (2018) reported on the different types of propagative budding such as stolonic budding, vascular budding, palleal budding, and pyloric budding. Such propagative budding types are detailed below.

Stolonial budding: In this mode of asexual reproduction, the buds are formed from a branching stolon. This method is not common in ascidians and is seen only in the species of *Perophora* and *Ecteinascidia* of the families Perophoridae (e.g., *Perophora viridis* and *Perophora multiclathrata* (=*Perophora orientalis*)); and Clavelinidae (e.g., *Clavelina lepadiformis* (Berrill, 1935; Kurn et al., 2011)).

Vascular budding: This type of budding is seen in some botryllid ascidians such as *Botrylloides leachi*, *B. schlosseri* var. *polycyclus* of the family Styelidae (Berrill, 1935). In this type of asexual reproduction, buds are known to arise from aggregations of hemocytes (lymphocyte-like cells) in tunic blood vessels.

Palleal or peribranchial budding: This type of budding is common in styelid ascidians (e.g., *B. schlosseri*, *B. leachi*, *B. violaceus*, and *Polyandrocarpa misakiensis*) (Kurn et al., 2011). In this budding process, buds are derived from evagination of the mantle (i.e., the body wall) at the sides of the body. In the above genera, three blastogenetic generations, namely, adult zooids with open siphons, buds, and budlets are seen in a colony (budlets are however derived from the evagination of the bud body wall). Further, in this mode of asexual reproduction, adult zooids close their siphons at regular intervals and are progressively resorbed in a process called takeover, while the buds attain their maturity and open their siphons. As a zooid yields more than one bud, colonies enlarge progressively.

Budding by strobilation: This type is common in the polyclinid and polycitorid ascidians. During this process, fragmentation of the elongated body into two or more parts takes place and each part undergoes partial regression and dedifferentiation followed by reorganization in order to form a complete individual. This type of budding has been reported in the species of Polyclinidae family (e.g., *Aplidium pallidum*, *Amoroucium yamazii*, and *Amoroucium constellatum* (Kurn et al., 2011).

Pyloric budding: It is typical of the didemnid ascidians (e.g., *Diplosoma listerianum* var. *gelatinosum*). In this type, two buds arise from esophagus and abdomen, respectively.

While the esophageal bud develops into a new lower part of the body, the abdominal bud becomes the new upper part. These two buds may unite together to form one zooid; or two zooids may be formed in which each will have an old and a new part.

1. *Diazona violacea* 2. *Aplidium elegans*

3. *Archidistoma aggregatum* 4. *Cystodytes lobatus* 5. *Nephtheis fascicularis*

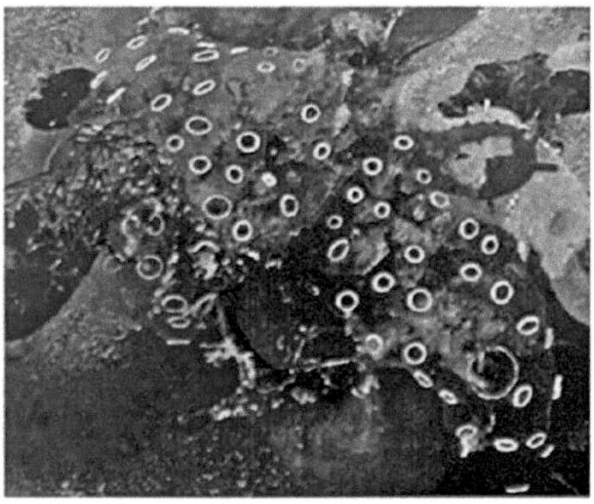
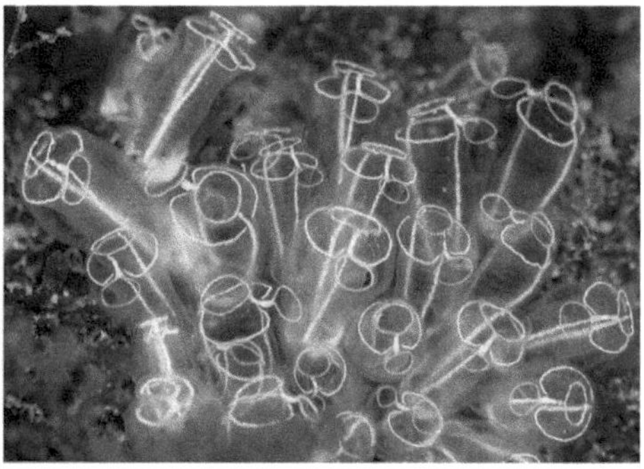

6. *Diplosoma listerianum* 7. *Clavelina lepadiformis*

8. *Clavelina picta*

9. *Chondrostachys cylindrica*

10. *Perophora listeri*

11. *Botrylloides violaceus*

12. *Botrylloides leachi*

13. *Botryllus schlosseri*

Images credit:

1. Matthieu Sontag. https://creativecommons.org/licenses/by/3.0/
2. Diego Delso. https://creativecommons.org/licenses/by-sa/4.0/
3. Bernard Picton. https://creativecommons.org/licenses/by/4.0/
4. Steve Lonhart. Public domain.
5. Nhobgood Nick Hobgood. https://creativecommons.org/licenses/by-sa/3.0/
6. Parent Géry. https://creativecommons.org/licenses/by-sa/3.0/
7. Diego Delso. https://creativecommons.org/licenses/by-sa/4.0/
8. Nhobgood Nick Hobgood. https://creativecommons.org/licenses/by-sa/3.0/
9. Peter Southwood. https://creativecommons.org/licenses/by-sa/3.0/
10. Bernard Picton. https://creativecommons.org/licenses/by-sa/4.0/
11. Frédéric André. https://creativecommons.org/licenses/by-sa/4.0/
12. Massimiliano DE MARTINO. https://creativecommons.org/licenses/by-sa/3.0/
13. Massimiliano DE MARTINO. https://creativecommons.org/licenses/by-sa/3.0/

Survival budding: The species of colonial ascidians possessing survival budding include *Aplidium multiplicatum*, *Botrylloides gascoi*, *Botrylloides leachi*, *Clavelina lepadiformis*, *Diazona violacea*, *Distomus variolosus*, *Eudistoma ritteri*, *Perophora* sp., and *Stolonica socialis* (Nakauchi, 1982)

CHAPTER 3

ASCIDIANS AS A HUMAN FOOD SOURCE

The increasing demand for protein-rich food in developing and under-developed countries has made these countries to explore other protein sources from the seas and oceans. In this regard, the edible species of ascidians, with their high protein, glycogen, and minerals form one of the important and underutilized marine living renewable resources and they have a great potential to be recognized as a valuable seafood. Eating ascidians has been reported to reverse the signs of aging (https://neuro-sciencenews.com/aging-sea-squirts-20570/). Among the tunicate fauna, solitary stolidobranch ascidians of the Pyuridae and Styelidae have been reported to be edible and are cultured or harvested as seafood.

Such species are *Microcosmus sabatieri*, *Microcosmus vulgaris*, *Microcosmus hartmeyeri*, *Polycarpa pomaria*, *H. aurantium*, *H. roretzi*, *Microcosmus polymorphus*, *Pyura chilensis*, *Pyura michaelseni*, *P. pachydermatina*, *S. clava*, and *Styela plicata*. In countries like Chile, Mediterranean, and Asia, these species are eaten raw, dried, cooked, or pickled (Gao et al., 2023).

Most of these species are sold in seafood markets as fresh or dried. *H. roretzi* is eaten raw and is commonly found in dishes of Japan and Korea where it is known as hoya and meongge, respectively. *S. clava* is a fresh, raw delicacy to Asian communities in the United States and Korea and is also serving as an aphrodisiac in Southern Korea. *S. plicata* is eaten fresh in both Korea and the Mediterranean and is also exported as a frozen product. Besides serving as food supplements, these ascidians are also processed into several recipes for human consumption. These recipes include ascidian pickles, freeze-dried ascidians, frozen ascidians, canned ascidians, sake-steamed hoya, hoya sashimi, hoya sashimi with half shell, hoyan Pra (Japanese Nam

Biotechnological Potential of Ascidians: Wacky and Weird Sea Squirts. Ramasamy Santhanam (Author)

Pla), and ascidian ale (hoya ale or blue beer Europe and North America have been reported to import processed *P. chilensis*, *S. clava*, *S. plicata*, *H. aurantium*, and *H. roretzi*. *M. hartmeyeri* is consumed in Japan; and *M. sabatieri* and *M. vulgaris* in France, Italy, and Greece. Both *M. vulgaris* and *P. pomaria* are eaten in the Mediterranean area. While *P. pachydermatina* is a delicacy in New Zealand, *P. praeputialis* (=*Pyura stolonifera*) is used in alcoholic beverages in New Zealand (Lambert et al., 2016).

The main edible species of ascidians and the countries of harvesting are given in Table 3.1.

TABLE 3.1 Main Species of Edible Ascidians and the Harvesting Countries

Species	Harvesting Countries
Boltenia ovifera	Russia, New England, and eastern Canada
Halocynthia aurantium	North Russia, Korea
Halocynthia roretzi	Korea, Japan
Microcosmus exasperatus	Mediterranean
Microcosmus squamiger	Mediterranean
Microcosmus hartmeyeri	Japan
Microcosmus sabatieri	Mediterranean
Microcosmus vulgaris	Mediterranean
Polycarpa pomaria	Mediterranean
Pyura chilensis	Chile
Pyura pachydermatina	New Zealand
Pyura praeputialis	Chile, Australia
Pyura vittata	Korea, Japan
Styela clava	Korea
Styela plicata	Korea, Mediterranean, southeast and southwest US, India

Source: Adapted from Lambert et al. (2016)

3.1 NUTRIENTS IN ASCIDIANS

Several species of large solitary ascidians have been reported to possess high nutritive value and they could serve as potentially healthy seafood high in protein and low in calories. For example, *H. roretzi* is known to contain several vitamins (C, E, and B12, in particular), minerals (sodium, potassium, phosphorus, iron, zinc, copper, calcium, and magnesium), and other components such as amino acids, cholesterol, fatty acid, pantothenic acid, and folic acid. *H. aurantium* is freeze-dried and made into food supplement tablets. Dried *H. roretzi* has been reported to possess antioxidant properties and is sold as health supplements in Korea at a price of US$270 for a box of 60-g packets. The levels of carbohydrate, protein, and lipid contents in several solitary and colonial species make these ascidians potential food sources (Lambert et al., 2016).

The proximate composition and nutritional values and health-promoting fatty acids of ascidians are given in Tables 3.2–3.4.

TABLE 3.2 Proximate Composition of Ascidians

Content	Range (%) of Values	Species with Minimum Value	Species with Maximum Value
Ash	0.14–1.56	*Didemnum moseleyi*	*Microcosmus. hellari*
Water	69.2–91.78	*Microcosmus. exasperatus*	*Didemnum psammatode*
Protein	3.8–20.01	*Didemnum psammatodes*	*Microcosmus. exasperatus*
Carbohydrate	2.2–8.29	*Policlinum indicum*	*Microcosmus. exasperatus*
Lipids	1.05–2.97	*Policlinum indicum*	*Microcosmus. exasperatus*

Source: Adapted from Ananthan et al. (2012).

TABLE 3.3 Nutritional Values of Ascidians

Component	Per 100 g Serving
Carbohydrates	Nil
Proteins	3.80 g
Fat	0.50 g
Saturated fatty acids	0.100 g
Monounsaturated fatty acids	0.120 g
Polyunsaturated fatty acids	0.200 g
Cholesterol (Sterol)	7.00 mg
Riboflavin (Vitamin B2)	0.200 mg
Minerals—Calcium, Ca	47.00 mg
Iron, Fe	3.30 mg
Potassium, K	102.00 mg
Sodium, Na	656.00 mg
Other ingredients—water	90.40 g
Ash	3.10 g

Source: Adapted from https://fit4art.com/en/blog/7051/Ascidians-nutritional-value-and-information

TABLE 3.4 Fatty Acid Profile (% of Total Fatty Acids) of Ascidians

Fatty Acid	%
14.0	1.45
16.0	12.56
16:1n-9	5.37
18:0	5.89
18:1n-7+n-9	11.98
18:2n-6	2.26
18:3n-3	2.38
18:4n-3	3.61
20:4n-6	3.11
20:5n-3	20.44
22:6n-3	11.85
∑n-3	40.07
∑n-6	6.94
∑n-3/∑n-6	5.79
∑SFA	22.39
∑MUFA	19.95
∑PUFA	48.48

Notes: ∑SFA: 14:0, 15:0, 16:0, 17:0; 18:0, 20:0, 21:0, 22:0, 24:0; ∑MUFA: 15:1, 16:1, 16:1n-7, 16:1n-9, 17:1, 17:1n-9, 18:1n-7+n-9, 20:1, 20:1n-9, 20:1n-7, 22:1n-11, 22:1n-9, 24:1n-9; ∑PUFA: 16:2, 16:2n-6, 16:3n-3, 16:4n-3, 18:2, 18:2n-6, 18:3n-6, 18:3n-3, 18:4n-3, 20:2, 20:2n-6, 20:3n-6, 20:3n-3, 20:4n-6, 20:4n-3, 20:5n-3, 22:4, 22:4, 22:5n-6, 22:5n-3, 22:6n-3.

Source: Adapted from Marques et al. (2021).

3.2 ASCIDIAN AS FOOD SUPPLEMENT

The sea squirt supplementation in the food of human beings has been reported to serve as a natural remedy to reduce inflammation and improve blood flow owing to the presence of its essential vitamins and minerals, and high-quality marine proteins and EPA and DHA, the omega-3 fatty acids. It is also amazing to note that the sea pineapple *H. roretzi* has more omega-3 fatty acids (DHA and EPA) than that of chicken meat (https://mybrainsupport.com/sea-pineapple/). Further, the sea squirt supplement is safe to consume with minimal side effects and it is known to provide a number of health benefits, especially in the treatment of arthritis, fibromyalgia, and other inflammatory diseases. Furthermore, *H. roretzi* is considered to be the superior source of plasmalogen which is a brain vitamin and supplement. Plasmalogen has been shown to improve memory, cognition, and hair health in laboratory mice models. Anon. (https://mybrainsupport.com/sea-pineapple/) reported that the sea pineapples and human brains are very similar in molecular composition. The commercial product, Ceregain, has been reported to be the best source of plasmalogen and is presently available in the United States (https://adoptandshop.org/the-invasive-nature-of-sea-squirts-understanding-their-biology-and-environmental-impacts/) where it is believed to be a household name. Experiments conducted with the Japanese adult volunteers who consumed active food containing ascidian-derived plasmalogen exhibited a marvelous score in composite memory which was based on verbal and visual memory, compared to the volunteers who consumed the placebo food (Watanabe et al., 2020).

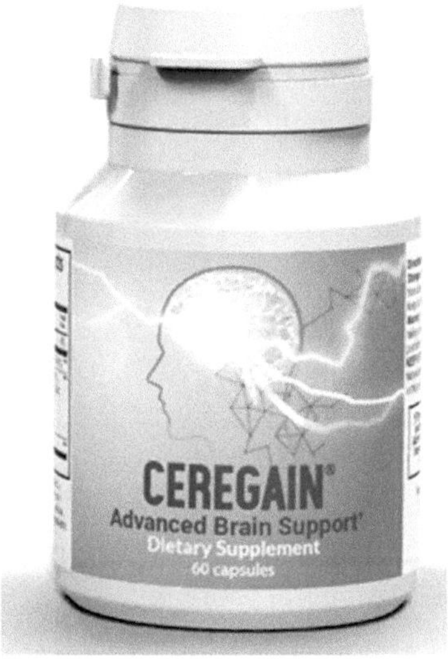

Image credit: Reprinted with permission from CEREGAIN.

3.3 BIOLOGY AND ECOLOGY OF EDIBLE ASCIDIANS

Didemnum moseleyi

Order: Aplousobranchia; family: Didemnidae

Image credit: Reprinted with permission from Jungle dragon.

Common name(s): Moseley's ascidian, white colonial tunicate; strawberry tunicate

Global distribution: Tropical species; Indian and Indo-Pacific Ocean and Red Sea

Ecology: Colony is a thin encrusting sheet and is commonly found on coral reefs and in lagoons at a depth range of 8–10 m. It may be an invasive species as it grows on top of sponges, coral, and other larger tunicates.

Biology: It resembles a red raspberry in color and its tunic "cormus" is covered with small holes, the zooids which are grouped around a common cloacal siphon. Each zooid has its own mouth, digestive, heart, and reproductive siphon. Spicules of 0.05 mm diameter are diverse and are crowded throughout the colony. It is a filter feeder, feeding on planktonic microorganisms.

Nutritional values: It is an edible species (Ananthan et al., 2012).

Didemnum psammathodes

Image not available

Common name(s): Chocolate tunicate

Global distribution: Though it is a native species of Northern Australia, it has a wide distribution in the Indo-West Pacific.

Ecology: It is found distributed on rocks, mangroves, coral reefs, ship hulls, and docks.

Biology: Tunic of this colonial tunicate is brownish-gray colored, smooth, and glossy.

Coloration of the colony, however, varies with the substrate and is ranging from cream in coral reef areas to brown in muddy environments. Brown coloration of the colony is due to the dense fecal pellets packed throughout the tunic. Colonies form thin encrusting sheets which may produce either irregular twig-like branches or fleshy lobes the size of which may be up to 8 cm long and 1 cm diameter green microalgae are also often embedded in the surface of the colony and its spicules (small crystalline structures) are seen embedded in the tunic and are scattered throughout. The presence of constricted common cloacal cavities in the colonies is a characteristic feature.

Zooids of the colony are less than 1 mm long, with fairly a wide atrial aperture. In the first row of the branchial sac, there are eight stigmata. This suspension feeder feeds on phytoplankton and detritus.

Nutritional values: It is an edible species (Ananthan et al., 2012). This species has been reported to possess amino acids and fatty acids of nutritional importance (Tables 3.5–3.8).

TABLE 3.5 Amino Acid Contents of *Didemnum psammathodes*

Amino Acid	mg/g
Leucine	540.9
Arginine	401.2
Lysine	385.4
Threonine	312.5
Isoleucine	254.1
Histidine	177.4
Methionine	175.6
Tyrosine	156.7
Valine	124.6
Phenylalanine	97.8
Alanine	72.1
Aspartate	71.2
Glutamate	24.6
Glycine	21.3
Serine	<0.1

Source: Adapted from Kumaran and Bragadeeswaran (2014).

TABLE 3.6 Monounsaturated Fatty Acids of *Didemnum psammathodes*

Fatty Acid	**%**
Cis-3 myristoleic acid	0.16
Trans-5 myristoleic acid	0.74
Cis-7 myristoleic acid	0.94
Cis-6-pentadecenoic	1.33
Cis-5-palmitoleic acid	1.06
Cis-6-palmitoleic acid	7.12
Trans-7-palmitoleic acid	0.71
Trans-9-palmitoleic acid	1.33
Cis-7-heptadecenoic acid	0.34
Trans-8-heptadecenoic acid	0.55
Cis-5-octadecenoic acid	0.11
Cis-7-octadecenoic acid	0.56
Oleic acid	6.13
Nonadecenoic acid	0.74
Cis-5-eicosenoic acid	0.13
Cis-7-eicosenoic acid	0.12
Trans-9-eicosenoic acid	0.31
Trans-7-docosenoic acid	0.22
Cis-9-docosenoic acid	0.44
Cis-3-tetracosenoic acid	0.72
Cis-6-tetracosenoic acid	0.84
Trans-9-tetracosenoic acid	0.92
Σ of MUFAs	25.52

Source: Adapted from Kumaran and Bragadeeswaran (2014).

TABLE 3.7 Polyunsaturated Fatty Acids of *Didemnum psammathodes*

Fatty Acid	**%**
Hexadecenoic acid	0.84
Trans-3-linoleic acid	0.63
Linoleic acid	5.96
Alfalinolenic acid	0.82

TABLE 3.7 *(Continued)*

Fatty Acid	%
Gammalinoleinic acid	1.33
Stearidonic acid	0.84
Octadecenoic acid	0.64
Eicosadienoic acid	0.84
Dihomogammalinoleic acid	0.33
Arachidonic acid	0.14
Eicosapentaenoic acid	10.34
Cis-6-eicosapentaenoic	0.36
Docosatrienoic acid	0.82
Docosatetraenoic acid	0.72
Docosahexaenoic acid	8.14
Σ of PUFAs	33.59

Source: Adapted from Kumaran and Bragadeeswaran (2014).

TABLE 3.8 Saturated Fatty Acids of *Didemnum psammathodes*

Fatty Acid	%
Lauric acid	0.64
Tridecyclic acid	0.71
Myristic acid	11.91
Pentadecyclic acid	1.92
Palmitic acid	11.63
Margaric acid	1.52
Stearic acid	4.31
Nonadecyclic acid	0.82
Arachidic acid	0.74
Heneicosanoic acid	2.16
Pehinic acid	2.92
Tricosanoic acid	1.94
Lignoceric acid	1.74
Σ of SFAs	32.96

Source: Adapted from Kumaran and Bragadeeswaran (2014).

Lissoclinum fragile

Image credit: Bernard Picton . https://creativecommons.org/licenses/by/4.0/

Common name(s): Caribbean ascidian

Global distribution: Tropical and subtropical Indo-Pacific and Western Central Atlantic.

Ecology: This sessile species is widely distributed in habitats such as mangrove roots, offshore oil platforms, stones, and pilings at a depth range of 0–10 m.

Biology: It is a thin, encrusting, and snowy white colony, the test of which is very fragile due to its dense, white, and calcareous burr-like spicules of ~0.02 mm diameter and its extensive shared cloacal cavities. Zooids of the colony are orange and are about 1.5 mm long. The branchial aperture is six-lobed and the branchial sac is large with four rows of stigmata. The intestinal loop is small, and the stomach is thin-walled. Two pear-shaped testes are seen in the intestinal loop and are united by a common duct. In this species, the presence of only a small group of eggs represents the ovary. Colonies of this species can reach up to 80 mm and 2–3 mm thick. It is a suspension feeder on phytoplankton and detritus.

Nutritional values: It is an edible species (Ananthan et al., 2012).

Eudistoma viride

Eudistoma sp.

Image credit: Bernard DUPONT. https://creativecommons.org/licenses/by-sa/2.0/

Family: Polycitoridae

Common name(s): Colonial ascidian

Global distribution: Tropical Western Central Pacific

Ecology: This sessile species is seen in marine lakes and on mangrove roots at a depth range of 0–3 m.

Biology: Colonies of this species are in the form of are bushes of small lobes. Zooids possess a yellow-green pigment in their body wall and are arranged as one to three systems in each lobe. Tunic is translucent. Both the siphons are shortened. Six oral lobes and 6–8 cloacal lobes are seen. In the branchial sac, there are about 25 stigmata arranged in a half row. The gut loop is fairly long and is provided with a posterior stomach in trapezium. Gonads are situated in the posterior gut loop.

Nutritional values: It is an edible species (Kumaran and Bragadeeswaran, 2014). This species has been reported to possess amino acids and fatty acids of nutritional importance (Tables 3.9–3.12).

TABLE 3.9 Amino Acid Contents of *Eudistoma viride*

Amino Acid	mg/g
Leucine	582.3
Arginine	365.4
Lysine	344.5
Threonine	295.6
Isoleucine	231.2
Histidine	168.4
Methionine	195.7
Tyrosine	124.3
Valine	140.5
Phenylalanine	105.2
Alanine	89.5
Aspartate	56.8
Glutamate	21.8
Glycine	<0.1
Serine	<0.1

Source: Adapted from Kumaran and Bragadeeswaran (2014).

TABLE 3.10 Monounsaturated Fatty Acids of *Eudistoma viride*

Fatty Acid	%
Cis-3-myristoleic acid	0.87
Cis-7-myristoleic acid	0.96
Cis-6-pentadecenoic acid	1.96
Cis-5-palmitoleic acid	1.01
Cis-6-palmitoleic acid	6.13
Trans-7-palmitoleic acid	0.18
Trans-9-palmitoleic acid	1.01
Cis-7-heptadecenoic acid	0.55
Trans-8-heptadecenoic acid	0.61
Cis-5-octadecenoic acid	0.18
Cis-7-octadecenoic acid	0.44
Oleic acid	7.12
Nonadecenoic acid	0.92
Cis-6-eicosenoic acid	0.16
Trans-9-eicosenoic acid	0.19
Trans-7-docosenoic acid	0.16
Cis-9-docosenoic acid	0.72
Cis-6-tetracosenoic acid	0.64
Trans-9-tetracosenoic acid	0.74
Σ of MUFAs	24.55

Source: Adapted from Kumaran and Bragadeeswaran (2014).

TABLE 3.11 Polyunsaturated Fatty Acids of *Eudistoma viride*

Fatty Acid	%
Hexadecenoic acid	0.13
Trans-3-linoleic acid	0.19
Linoleic acid	5.11
Alfalinolenic acid	0.92
Gammalinoleinic acid	1.01
Stearidonic acid	0.93
Octadecenoic acid	0.11
Eicosadienoic acid	0.14
Eicosapentaenoic acid	11.10
Cis-6-eicosapentaenoic	0.11
Docosatrienoic acid	0.74
Docosatetraenoic acid	0.63
Decosapentaenoic acid	0.83
Docosahexaenoic acid	7.12
Σ of PUFAs	29.07

Source: Adapted from Kumaran and Bragadeeswaran (2014).

TABLE 3.12 Saturated Fatty Acids of *Eudistoma viride*

Fatty Acid	%
Lauric acid	0.16
Tridecyclic acid	0.18
Myristic acid	1.61
Pentadecyclic acid	1.43
Palmitic acid	10.72
Margaric acid	1.94
Stearic acid	6.22
Nonadecyclic acid	2.51
Arachidic acid	1.55
Heneicosanoic acid	1.01
Pehinic acid	1.22
Tricosanoic acid	1.55
Lignoceric acid	1.61
Σ of SFAs	31.71

Source: Adapted from Kumaran and Bragadeeswaran (2014).

Family: Polyclinidae

Polyclinum constellatum

Image not available

Common name(s): Red Sea tunicate

Global distribution: This species has a pantropical distribution and it is common throughout the Caribbean (and warm waters).

Ecology: This sessile, colonial encrusting tunicate occurs in mangrove roots, rocks, and dead corals. It is also known to overgrow cultured mussels and in man-made structures such as wharfs, buoys, pilings, and dock floats. These animals may also be associated with cnidarian hydroids or other epifaunal components. In these colonial animals, only a small part of their base is found attached to the substrata.

Biology: It is gray, purplish brown, gray, or green in coloration. White or beige systems of zooids are seen. Colonies are firm and cartilaginous with 2.5–6.5 cm long and 0.5–2.0 cm thick. Zooids are 5–7 mm long when they are straightened; and about 20 zooids are arranged in circular systems. There are oral siphons which are fringed with six triangular lobes. Reproductive organs are present in the postabdomen. It is a suspension feeder on algae and detritus.

Nutritional values: It is an edible species (Ananthan et al., 2012).

Polyclinum indicum

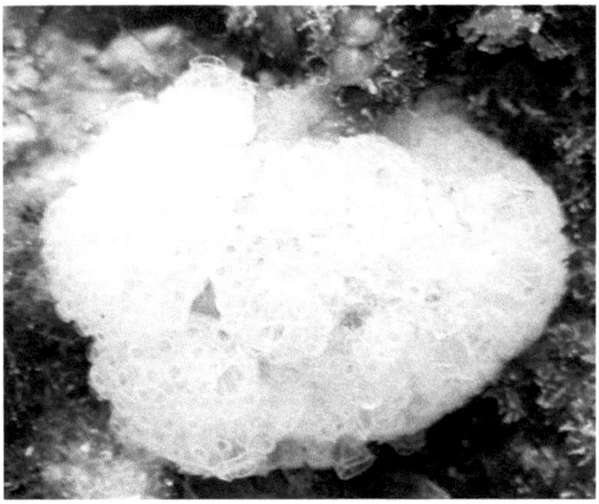

Polyclinum sp.

Image credit: Matthieu Sontag. https://creativecommons.org/licenses/by-sa/3.0/

Common name(s): Not designated

Global distribution: Indian Ocean

Ecology: It is found distributed in estuaries and rocky shores. Colonies are found attached to the bottom of boulders and stones at the level of the tides.

Biology: Colonies of this species vary in shape as the younger ones are oval or pear-shaped with narrow base of attachment and round upper exposed surface. On the other hand, fully grown colonies which have a maximum diameter of 6 cm and height 5 cm possess umbrella-shaped upper exposed surface with a broader base of attachment. Common cloacal openings are found scattered on the exposed surface. Surface of the colonies are invariably encrusted with a thin layer of sand. Color of the colonies is light brown or pale red and rarely dull green. Zooids are arranged in systems of about 20–40 and three or more such systems are seen around one common cloacal opening. Branchial openings possess a whitish color on their margins. Based on the coloration of the colony, the tunic is transparent with green or red tinge.

Nutritional values: It is an edible species (Ananthan et al., 2012).

Polyclinum madrasensis

Image not available

Common name(s): Not designated as it is a new species.

Global distribution: Southeast coast of India

Ecology: It occurs in shallow benthic areas.

Biology: Not available

Nutritional values: It is an edible species (Ananthan et al., 2012).

Order: Stolidobranchia; family: Pyuridae

Boltenia ovifera

Image credit: A. C. Tatarinov. https://creativecommons.org/publicdomain/zero/1.0/

Common name(s): Sea onion, sea potato, stalked tunicate

Global distribution: It is native to the Bering Sea and is found in the Arctic to the South of Cape Cod.

Ecology: This solitary, sessile species occurs in subtidal rocky bottoms and tidepools. It is found invariably anchored in kelp holdfasts.

Biology: Color of the animal is orange, reddish, just brown, or tan. It is found to grow on a stalk which is more than three times the length of the animal. As other tunicates, it is also a filter feeder feeding on phytoplankton size of the animal may vary from 7.5 to 30 cm.

Nutritional values: This edible species as a subsistence food resource called "Oopa" for the native communities in the Bering Strait region during holiday celebrations. It is added to stews, frozen whole, and thinly sliced and consumed with crystalized briny water from inside or blanched and peeled before consumption (Bugaj, 2022).

Halocynthia aurantium

Image credit: Reprinted with permission from Alexander Semenov.

Common name(s): Pacific Sea Peach, sea peach, ice floe tunicate

Global distribution: This polar and subtropical species is native to the Bering Sea and is found in the Arctic and Northern Pacific; and in the Southern, Eastern Sea of Korea, and Northern Sea of Japan.

Ecology: It is found on rocks in subtidal areas at depths of about 100 m. Bod of the animal is found fixed directly to the substrata.

Biology: This solitary, sessile ascidian has a maximum length of 18.0 cm and it is typically barrel shaped with red or orange coloration. Its tunic is smooth or wrinkled. At the top of the animal, there are two siphons. It is known to be the major food source of crabs and sea stars.

Nutritional values: It is an edible species and is consumed in the countries of East Asia, including Korea and Japan. This edible species as a subsistence food resource called Oopa for the native communities in the Bering Strait region during holiday celebrations. It is added to stews, frozen whole and thinly sliced and consumed with crystalized briny water from inside or blanched and peeled before consumption (Bugaj, 2022). Kononova et al. (2022) reported that a valuable food supplement "HAURANTINUM" is derived from the tunic of this species as an alcohol extract. This extract contains various biologically active substances such as free amino acids, fatty acids, phospholipids, neutral lipids, and prostaglandins. Apart from these, several micro- and macro-elements, including vanadium have also been reported from this extract.

Halocynthia roretzi

Image credit: bifyu. https://creativecommons.org/licenses/by-sa/2.0/

Common name(s): Sea pineapple, Korean common sea squirt; Japanese sea squirt

Global distribution: Subtropical Northwest Pacific: Japan.

Ecology: This benthic, sessile species lives in shallow water and is usually attached to rocks, stones, and artificial structures as fouling organisms. It is highly adapted to cold water surviving in water temperatures between 2 and 24°C (36–75°F).

Biology: It is a solitary, soft-bodied and bilaterally symmetrical organism. Individuals of this species possess an ovoid body which is covered with large conical bumps. Apically, the body has two siphons which are used as water inlet and outlet for filter-feeding. Basally, it bears many branching roots that are used in attachment to rocks. Based on spawning, sea pineapples are of three types, namely, A, B, and C. Type A possesses long papillae (tunic protrusions) and its tunic coloration is light pink. This type is known to spawn in morning hours in November. The brownish red or reddish orange type B spawns in evening or night during October–November. Type C is bright red and it has short thick papillae and this type spawns mainly in the afternoon during April (Numakunai and Hoshino, 1973, 1974). The tunic in live sea pineapple is made of thick bundles of tunic fibers that interface to each other.

Nutritional values: The total combined amino acid contents of the muscle of the wild sea pineapple have been reported to be 11,595.4 mg%; and the major amino acids were Asp, Glu, and Lys. Further, the major fatty acids of this species were 14:0, 16:0, 16:1n7, 18:1n7, 18:4n3, 20:5n3, and 22:6n3, and composition ratio of n3 PUFAs was 47.0% (Oh et al., 1997). Furthermore, the edible parts of this species have been reported to possess several trace elements like Fe, Mn, Cu, Zn, Ca, Se, Cr, Sn, Li, Ni As, Pb, Cd, and Sr. Among these trace elements, the concentration of Zn is much higher compared to other marine animals. Further, it is worthy of mention ere that level of toxic heavy metals in these animals is much lower than the level prescribed by the national standards thereby making this species safe to consume (Gao et al., 2023). Nam-Do et al. (2014) and Li et al. (2007) reported on the nutritional composition of sea pineapple and the percentage values of the different constituents are given in Table 3.13.

Oh et al. (1997) reported that *H. roretzi* specimens collected by them showed 77.5% moisture, 2.5% ash, 6.6% glycogen, 11.3% crude protein, and 1.1% crude lipids. Among the lipids, neutral lipids constituted 83%, and phospholipids formed the remaining (17%). Among the amino acids of this species, Asp, Glu, and Lys were the major ones; and major fatty acids were found to be 14:0, 16:0, 16:1ω7, 18:1ω7, 18:4ω3, 20:5ω3, and 22:6ω3. The ω-3 PUFAs constituted 39%. The major inorganic ingredients of this species were found to be Na^+, K^+, Cl^-, and PO_4^3. Zhao and Li (2016) resorted that the animals of *H. roretzi* weighing 70.4 g exhibited 46.52% carbohydrates, 0.28% lipids, 38.08% protein, and 11.55% ash. The nutritional profile of this species is comparable to certain seafood products thereby making it highly suitable for human food applications.

TABLE 3.13 Nutritional Composition of *Halocynthia roretzi* Meat

Content	%[a]	%[b]
Moisture	88.7	–
Ash	1.4	25.8
Carbohydrate	3.0	9.2
Protein	6.0	39.6
Lipid	0.8	6.5
Amino acids (total) (TA)	–	29.5
Essential amino acids of TA	–	51.4
Delicious amino acids of TA	–	36.4
Essential amino acids	–	90.0
Minerals (Na, Mg, Ca, P, Fe, Zn, Cu, Co, V, and Mn)	–	ND

Source: Adapted from [a]Nam-Do et al. (2014); [b]Li et al. (2007).

Culinary uses: Though there are several types of sea pineapples available in the world, only a few countries such as Japan, Korea, Chile, and France eat sea pineapples. In Korea, it is mostly eaten raw with vinegared hot pepper paste (gochujang). However, it is also used to add flavor to kimchi or pickled. In Japan also, the sea pineapple is most commonly eaten raw. While eating, the animal is sliced vertically and its internal organs are removed. It is then served with vinegared soy sauce. This species is also occasionally deep-fried, grilled (https://www.umai-aomori.com/202106/sea-pineapple/). The raw eating of sea pineapple is due to its peculiar flavor resembling rubber dipped in ammonia, which is due to the presence of an unsaturated alcohol called cynthiaol (https://en.wikipedia.org/wiki/Cynthiaol).

Meongge–hoe
(unprocessed sea pineapples)
Image credit: Andrew Bell. https://creativecommons.org/licenses/by/2.0/

Meongge–hoe–bibimbap
(raw sea pineapples containing rice)
Image credit: Junho Jung, https://creativecommons.org/licenses/by-sa/4.0/

(Mixed rice with raw sea pineapples)

Image credit: Jpatokal, https://creativecommons.org/licenses/by-sa/4.0/

In Japan, the sea pineapple is served as sashimi in restaurants.

Medicinal values: Water extracts derived from the edible portions of this species have been reported to possess anticancer properties by significantly inhibiting the growth of HepG-2 (human liver cancer cells). Further, this species has shown various kinds of bioactivities, such as antihypertension, antidiabetic, and immunity enhancement. This species may have a potential role as functional nutraceuticals or in chemotherapy as an adjunct (Zhu et al., 2022).

Herdmania pallida

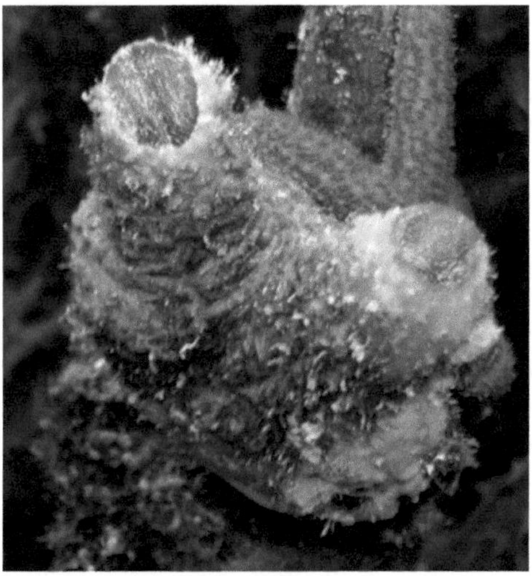

Image credit: Public domain

Common name(s): Not designated

Global distribution: Tropical West Atlantic and Indo-Pacific; Panama, Caribbean islands, Brazil

Ecology: It is found only on mangrove roots at a depth of 0.5–1.0 m.

Biology: Individuals of this species are 4.0–5.0 cm long and are usually covered by encrustations. These animals may be recognized by their characteristic siphons which have an iridescent bluish lining. The body of this species is more or less globular and is covered by a soft tunic. In live animals, gonads are orange; internal wall of siphons is blue and iridescent; and tunic is light pink. Needle-like calcareous spicules are seen on the body wall. Pharynx has eight to nine folds in its sides.

Nutritional values: It is high in protein (45 mg/100 mg), amino acids (1.201 mg/100 mg), lipid (18 mg/100 mg), and carbohydrates (4.8 mg/100 mg) and the flesh of this species may be used in the preparation of pickles (Lambert et al., 2016; Tamilselvi et al., 2010).

3.3.1 MICROCOSMUS

All species of this genus *Microcosmus* have been reported to be edible. However, only a few species of M*icrocosmus* such as *Microcosmus exasperates*, *M. hartmeyeri*, *Microcosmus helleri*, *M. polymorphus*, *M. sabatieri*,*Microcosmus squamiger*,and*M. vulgaris*(=*Microcosmus sulcatus*) are only reliably eaten Further, among these species, *M. polymorphus*, *M. sabatieri*, and *M. vulgaris* are harvested commercially in the Mediterranean even from the 1st century AD. Other species of *Microcosmus* which are believed to be eaten include *Microcosmus arenaceus*, *Microcosmus australis*, *Microcosmus albidus*, *Microcosmus anchylodeirus*, *Microcosmus anomalocarpus*, *Microcosmus bitunicatus*, *Microcosmus claudicans*, *Microcosmus curvus*, *Microcosmus hirsutus*, *Microcosmus longicloa*, *Microcosmus nudistigma*, *Microcosmus oligophyllus*, *Microcosmus planus*, *Microcosmus propinquus*, *Microcosmus vesiculosus* and *M. vulgaris (=Microcosmus sulcata)*, *Microcosmus senegalensis*, *Microcosmus stoloniferus*, *Microcosmus glacialis*, *Microcosmus hernius*, *Microcosmus madagascariensis*, *Microcosmus miniaceus*, *Microcosmus multiplicatus*, *Microcosmus multitentaculatus*, *Microcosmus tuberculatus*, *Microcosmus pacificus*, *Microcosmus psammiferus*, *Microcosmus pupa*, *Microcosmus santoensis*, *Microcosmus savignyi*, and *Microcosmus trigonimus* (https://notesfromdisgraceland.wordpress.com/hobbies/food/).

Microcosmus exasperatus

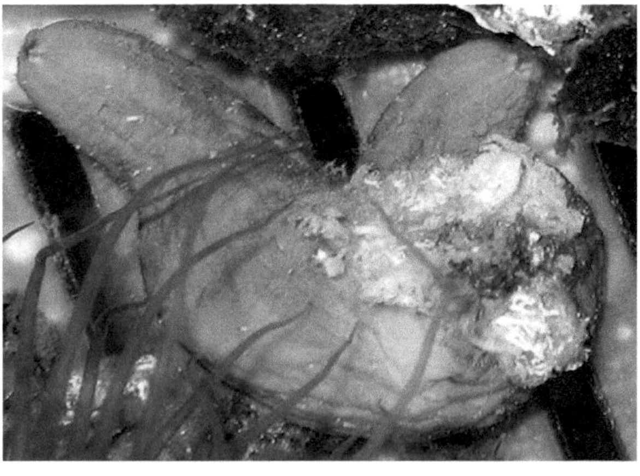

Source: Ascidians. http://www.ascidians.com/families/pyuridae/Microcosmus_exasperatus/microcosmusexasperatus2.htm

Common name(s): Exasperating tunicate

Global distribution: Atlantic, Indo-Pacific, Caribbean Sea, Mediterranean Sea, Brazil

Ecology: This solitary tunicate commonly occurs in dense clumps or aggregates. It is attached to the substratum by its posterior ventral side. Dense populations and heavy epibiont coverage of this species may contribute to its establishment as an ecosystem engineer species.

Biology: Body of this species is ovate and irregular. Test is rough and leathery. Small sharp spines are seen at the margins of the oral siphon which has 12–25 branched tentacles. Dorsal lamina is smooth and 8–10 folds are seen on either side of the pharynx. Its elongated gonads are provided with 3–5 lobes. Internal siphonal spines are longer and pointed and are 40–50 μm long, and posteriorly hooked.

Nutritional values: This species is collected or cultivated for consumption in countries like Europe and Asia, and also Chile. The carbohydrate, protein, and lipid contents of this species have been reported to range from 14.1% to 11.2%; from 17.3% to 23.5%; and from 2.0% to 2.7%, respectively. Among the essential amino acids of this species, leucine showed the maximum values (567.3 mg); and among the nonessential amino acids, aspartic acid recorded the maximum value (0.212 mg). Further, the saturated fatty acids (SAFA), monounsaturated fatty acids (MUFA), and PUFAs were found to range from 0.110 to 1.664 mg; from 1.034 to 1.565 mg; and from 1.293 to 1.786 mg, respectively (Karthikeyan et al., 2010). Meenakshi et al. (2012) reported on the vitamins A, D3, and K of this species with the values of 20, 75, and 15.7 μg/g, respectively.

Microcosmus hartmeyeri

Image not available

Common name(s): Not designated

Global distribution: Japan coast

Ecology: It occurs in rocky reef areas of about 10 m depth with a water temperature of 11°C.

Biology: It is known to grow a maximum length of 15 cm.

Nutritional values: It is eaten locally in Japan (Nanri et al., 1992)

Microcosmus helleri

Image credit: Public domain

Common name(s): Not designated

Global distribution: Tropical Indo-West Pacific and Atlantic Ocean: East Africa, Brazil, Caribbean Sea

Ecology: This sessile species is found living on loose sediments, weakly attached shells or sea weeds at a depth range of 58–60 m. Sand, algae, and sessile invertebrates have been reported to encrust these ascidians.

Biology: These are small, pale individuals of with a size range of 2.5–3.0 cm in length and 2–3 cm in width. Its siphons are apical and are far from each other. Four small triangular lobes are seen on both the siphons. Internal wall of the oral siphon has a thick velum and this siphon has four lobes. On the other hand, the atrial velum is delicate and thin and is also divided into lobes which are overlapping. In this species, there are no spines on the siphons. A total of 14–18 oral tentacles with two sizes are present. These tentacles are branched three times. Minute papillae cover the peripharyngeal area. It is horseshoe-shaped dorsal tubercle aperture has enrolled ends. Dorsal lamina is simple, continuous, and is larger toward the esophageal opening. Pharynx is with six folds on either side.

Nutritional values: It is an edible species (Ananthan et al., 2012).

Microcosmus polymorphus

Image not available

Common name(s): Not designated

Global distribution: It is widely distributed in the Eastern Mediterranean (Greek Aegean Sea) and Western Mediterranean Seas; and Atlantic coast of Portugal.

Ecology: It is known to occur chiefly in the subtidal rocky substrates, coarse sediments (rocks with gravel), and among *Posidonia oceanica* meadows at a depth range of 1–50 m.

Biology: This solitary ascidian has an irregular body and is 5–10 cm in size. Tunic of this species is thick and flexible, not wrinkled, and is sometimes covered with sand. Coloration of this tunic is red with yellowish-white streaks. Other specific characters of this species are short, blunt-pigmented dark red siphonal spines, and seven branchial folds at each side.

Nutritional values: It is eaten raw in countries like Italy and France. It is commercially harvested in the Mediterranean (Cihangir et al., 2011).

Microcosmus sabatieri

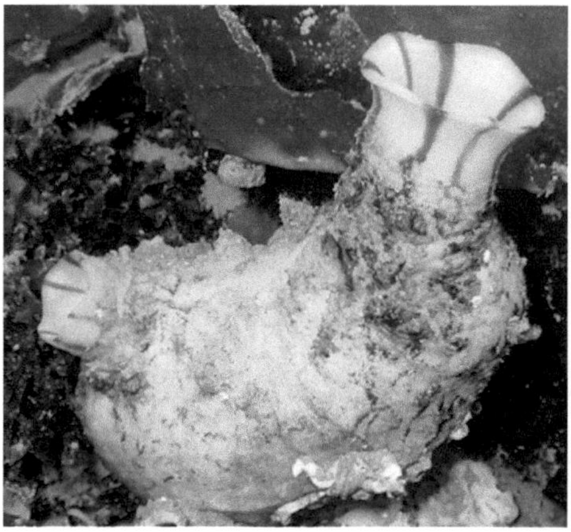

Image credit: Di Sylvain Le Bris. https://creativecommons.org/licenses/by-sa/4.0/

Common name(s): Sea violet, sea fig, or sea egg

Global distribution: Subtropical Atlantic and the Mediterranean Sea

Ecology: This sessile species occurs in shallow waters at a depth range of 15–32 m.

Biology: It has a rocky shape or large potato appearance with blackish-brown coloration on the outside. Its tunic is thick and tough and is covered with epibiotes (fixed organisms); hence, its Latin name *Microcosmus*. It takes in seawater through its top siphon and ejects it through a side siphon. As the water passes through, its branchial structure retains its microorganisms necessary for its nutrition, as well as oxygen. Size of the organisms ranges from 15 to 20 cm.

Nutritional values: Though all the species of *Microcosmus* are edible, *M. sabatieri* has been reported to be the most preferred food in parts of Southern Europe, such as Greece, France, Italy, Croatia, or Montenegro. It may be eaten raw or with other condiments such as vinegar or lemon juice with shallots. As it has a strong iodine taste, this species may not appeal to everyone. Its taste is very crisp and fresh with a distinct flavor resembling a freshly caught seafood. It is commercially harvested in the Mediterranean (Cihangir et al., 2011; https://notesfromdisgraceland.wordpress.com/hobbies/food/).

Microcosmus squamiger

Image credit: Reprinted with permission from Yu Lun-Pan.

Common name(s): Scaly tunicate

Global distribution: It is a native species of Australia and is found presently distributed Mediterranean Sea, the Atlantic coast of Spain, West Coast of the United States, Mexico, Canary Islands, and South Africa.

Ecology: It is a subtidal, rocky, epibenthic species, and it is found in the natural and artificial substrata.

Biology: It is a solitary tunicate with a leathery, tough, and purple tunic. It is usually found as aggregates or dense clumps with a maximum size of 50 mm diameter. Tunic is tough and leathery, hard or brittle, and is in purple coloration. Surface of the tunic of the animal has ridges and swellings with adhesive projections. Apertures are situated on short, wart-like siphons and are about one-third of the body length. In some specimens, unlike the atrial siphon, the oral siphon may be long and straight and terminal in position. The internal siphonal spines are very short, about 15–25 μm long, and fingernail shaped with serrated rims. Like other species of ascidians, it is a suspension feeder on phytoplankton and detritus.

Nutritional values: It is an edible species with rich nutritional values. The proximate composition, mineral and fatty acid contents, amino acid profile, and carotenoids of this species have been estimated and the percentage values are given in Tables 3.14–3.18.

TABLE 3.14 Proximate Composition of *Microcosmus squamiger*

Item	%
Moisture	88.9
Ash	36.2
Carbohydrate	40.8
Protein	53.9
Fat	5.3

Source: Adapted from Pan et al. (2021).

TABLE 3.15 Mineral Content of *Microcosmus squamiger*

Mineral	mg/g Dry Weight
Iron	4.42
Manganese	0.41
Aluminum	4.60
Selenium	ND
Molybdenum	ND
Lithium	ND
Vanadium	4.63
Sodium	36.24
Potassium	11.88
Calcium	14.29
Magnesium	6.75
Phosphorus	2.66

Note: ND, not detected.

Source: Adapted from Pan et al. (2021).

TABLE 3.16 Fatty Acid Content of the Methanol Extract of *Microcosmus squamiger*

Fatty Acid	μg/mg, dw
Oleic acid	0.33
Palmitoleic acid	1.52
Cis-10-pentadecenoic acid	ND
Myristoleic acid	ND
Cis-10-Heptadecenoic acid	ND
Elaidic acid	1.68
Erucic acid	NQ
Nervonic acid	ND
ΣMUFA	3.53
Eicosapentaenoic acid	2.39
Docosahexaenoic acid	0.51
Arachidonic acid	1.17
Linoleic acid	0.04
Linoelaidic acid	0.06
Eicosadienoic acid	0.94
γ-Linolenic acid	NQ
α-Linolenic acid	0.44
Homo-γ-linolenic acid	NQ
ΣPUFA	5.56

Note: ND, not detected; NQ, not quantified.

Source: Adapted from Pan et al. (2021).

TABLE 3.17 Amino Acid Profile of *Microcosmus squamiger*

Amino Acid	mg/g Dry Weight
Alanine	2.94
Cysteine	0.15
Tyrosine	0.33
Aspartic acid	0.92
Asparagine	0.67
Glutamic acid	2.41
Glutamine	0.88
Glycine	1.72
Proline	5.15
Serine	0.82
Arginine	1.65
Histidine	0.75
Lysine	1.29
Threonine	1.31
Isoleucine	0.17
Leucine	0.27
Valine	0.73

TABLE 3.17 *(Continued)*

Amino Acid	mg/g Dry Weight
Tryptophan	0.23
Methionine	0.85
Phenylalanine	0.25
Taurine	2.66
Ornithine	0.65

Source: Adapted from Pan et al. (2021).

TABLE 3.18 Carotenoid Content of the Dichloromethane Extract of *Microcosmus squamiger*

Compound	µg/g Dry Extract
Astaxanthin	0.03
Fucoxanthin	1.66
Anteraxanthin	NQ
Zeaxanthin	NQ
Lutein	0.14
Fucoxanthinol	2.51
Total	4.34

Note: NQ, not quantified.

Source: Adapted from Pan et al. (2021).

Microcosmus vulgaris (=Microcosmus sulcatus)

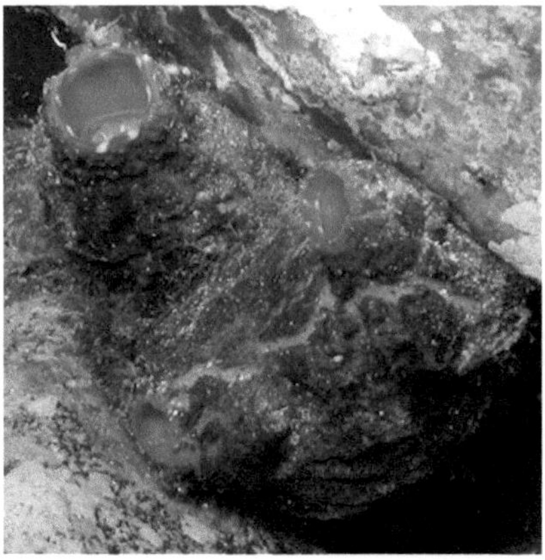

Image credit: Reprinted with permission from Mario Munaretto.

Common name(s): Violet, sea fig

Global distribution: Subtropical Western Indian Ocean, Atlantic, and the Mediterranean Sea

Ecology: This sessile species occurs at a depth range of 2–12 m.

Biology: It is relatively a large species with tawny red or brilliant red in coloration. Its hard tunic is with many folds. While its siphons are with four red lobes, branchial chamber is with seven folds on every side.

It is appreciated for its powerful "marine, iodized" flavor—trimethylamine was attributed to the other and defined as a key compound of sea fig aroma.

Nutritional values: This edible species is commercially harvested in the Mediterranean (Cihangir et al., 2011) and is eaten in Europe (Fulgione et al., 2006).

Essential trace elements: This species has been reported to contain essential trace elements in human nutrition such as iron (Fe), cobalt (Co), and zinc (Zn) at concentrations of 0.5×10^{-8}, 0.5×10^{-6}, and 1.3×10^{-5} g/g wet wt, respectively (Strohal et al. https://aslopubs.onlinelibrary.wiley.com/doi/pdf/10.4319/lo.1969.14.2.0265).

Toxicity: While consuming this species, precautionary steps have to be taken. In this case, the collection area is very important as organisms collected from polluted areas may be toxic. First, human intoxication due to consumption of PSP *M. vulgaris*, harvested from the Adriatic Sea has been reported by Roje-Busatto and Ujevi (2014). They reported on the presence of five paralytic shellfish toxins (PSP), namely, saxitoxin (STX), gonyautoxins 2 and 3 (GTX2,3), decarbamoylsaxitoxin (dcSTX), decarbamoylgonyautoxins 2 and 3 (dcGTX2,3), and gonyautoxin 5 (GTX5), the maximum concentrations of which were found to be 653.17, 78.11, 655.72, 389.18, and 585.2121 mg/kg, respectively.

Pyura chilensis

Image credit: sergio.majluf. https://creativecommons.org/licenses/by/2.0/

Common name(s): Red sea squirt, sea tulip

Global distribution: Chilean and Peruvian coasts

Ecology: This sessile species occurs in intertidal and subtidal habitats. Although this species is solitary, it is commonly found in highly aggregated clumps and is rarely isolated.

Biology: Individuals of this species are ovoid in shape measuring up to 5 cm in diameter. Color of the organisms is generally orange, but diverse tones of brown to almost black are also encountered.

Tunic is tough and its surface has wrinkles, protuberances, and papillae. Siphons are red-colored, quadrangular, and relatively close together. Internal tunic of the siphons has pointed spinules. Mantle has a strong musculature and it has orange turning red in the siphonal area. Oral siphon has more than 18 branched tentacles. Atrial siphon has a large bi-lobed velum. Branchial sac has six folds on each side.

Nutritional values: The mantle and internal structures of this species are edible. It is an important food of Chile (Manríquez and Castilla, 2007). It is believed to possess aphrodisiac properties (https://www.google.com/search?q=aphrodisiac+properties+of+Pyura+chilensis-+Wikipedia). Chile has been reported to export this species to several countries, including Japan and Sweden. As it is considered to be an important species in the economy of this country, it was celebrated on a special postage stamp.

Pyura pachydermatina

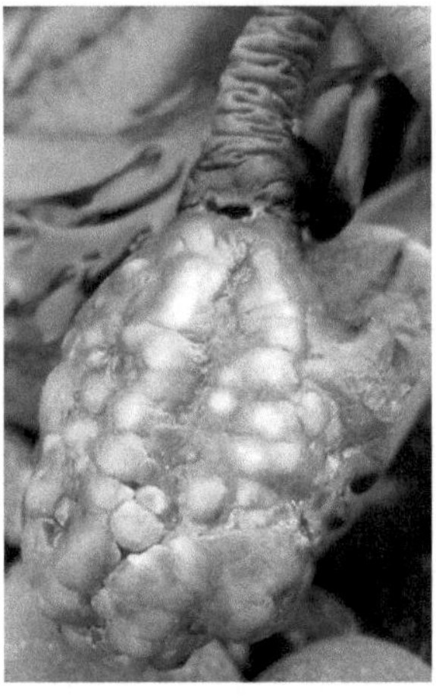

Image credit: wild_wind. https://creativecommons.org/licenses/by/4.0/

Common name(s): Kaeo sea tulip

Global distribution: It is native to the shallow seas around New Zealand in Pacific.

Ecology: This sessile organism is found attached to rocky substrata in the shallow, low intertidal to subtidal areas. It is known to prefer cooler waters.

Biology: It is the largest, solitary ascidian and it has a club-shaped body which is covered in a pink colored, leathery tunic. Coloration of the body is either off-white or with a shade of reddish-purple. It has a long stalk up to 40 cm in length which is attached to the rock. It can grow to more than a meter in length. It is a filter feeder and it has a complex gut system called pharyngeal basket which filters out the planktonic food organisms from the seawater.

Nutritional values: It is an edible and commercially harvested species (https://www.sealifebase.se/summary/Pyura-chilensis.html)

Pyura praeputialis *(=Pyura stolonifera)*

Image credit: Tesk0002. https:// creativecommons.org/licenses/by-sa/3.0/

Image credit: Marcriusvil. https:// creativecommons.org/licenses/by-sa/3.0/

Common name(s): Red bait

Global distribution: Tropical Indo-Pacific: Africa, Australasia, and South America.

Ecology: This sessile species commonly occurs on rocky intertidal reefs at a depth range of 10–12 m. It is an encrusting organism and is usually with a thick covering. Beds of this species provide an important habitat for many epiphytic organisms, such as algae, limpets, snails, and chitons which give this species a green-brown appearance.

Biology: It is a large, solitary barrel-shaped, tunicate reaching a maximum size of 3035 cm in height. Body of this species is actually orange or pink. It has a thick pale tunic which is wrinkled and brown in coloration. Its large exhalant and inhalent siphons project upward or sideways and are provided with slightly scalloped edges which are set close together.

Nutritional values: It is an edible species (Castilla and Defeo, 2001).

Pyura sacciformis *(=Pyura michaelseni)*

Image not available

Common name(s): Not designated

Global distribution: Australia, Moorea, Tahiti, Japan, and Fiji.

Ecology: It occurs in the intertidal rocky reefs at depths up to 80 m. It is found securely attached to the concrete and under the overhangs, and it is entirely covered in epibiota.

Biology: Body of this species is depressed dorso-ventrally and is attached to the substratum by its whole ventral side. The anterior part of the body is compressed more strongly than the posterior part. Tunic is thick and leathery and its surface is smooth and whitish except in the central portion on the dorsal side, where it is wrinkled around the two apertures. The inner surface is whitish and mantle is yellowish white. Mantle is thick on the dorsal side and is very thin on the ventral side. Siphons are situated at each one-third on the length of body and are tinted violet-brown on the tip. Branchial sac has six folds on each side. Like all ascidians, it is a filter feeder on phytoplankton and detritus.

Nutritional values: This edible species had total sugar, crude protein, and crude lipid at 2.70%, 12.30%, and 1.40%, respectively (Gao et al., 2023). The total combined amino acid contents of the muscle of theis wild sea pineapple have been reported to be 12,152.7 mg%, and the major amino acids were Asp, Glu, and Lys. Further, the major fatty acids of this species were 14:0, 16:0, 16:1n7, 18:1n7, 18:4n3, 20:5n3, and 22:6n3, and composition ratio of n3 PUFAs was 46.5% (Oh et al., 1997).

Family: Styelidae

Polycarpa pomaria

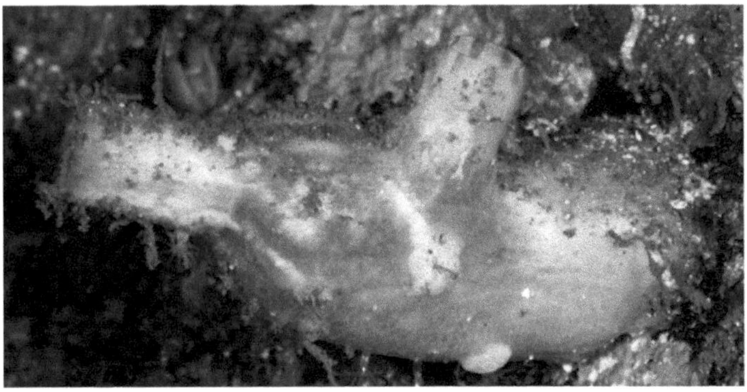

Image credit: Bernard Picton, https://creativecommons.org/licenses/by-sa/4.0/

Common name(s): Fruit sea squirt

Global distribution: It is native to the northeastern Atlantic Ocean

Ecology: This benthic species occurs at depths of about 500 m. It has been reported to show reef-stabilizing effect by joining pieces of coral colonies thereby conferring solidity and strength to the reef complex and its development.

Biology: It is a solitary tunicate with ovoid, conical, or globular shape and is growing to a maximum size of 7 cm long. Its oral siphon located at the apex of the animal is with four lobes and is encircled by a total of 56 tentacles of different sizes. Atrial siphon has four lobes. Tunic of this species is leathery and wrinkled and brown in coloration. It has several fleshy papillae. Internally, its body wall has a number of drop-shaped gonads (polycarps). Its pharynx wall is perforated by a number of bands of stigmata (slits).

Like other species of ascidians, it is known to feed by taking water in through its buccal siphon, filtering out its food organisms, namely, phytoplankton and sending out the water via its atrial siphon.

Nutritional values: It is an edible species in the Mediterranean region (Lambert et al., 2016).

Styela clava

Image credit: Matthieu Sontag. https://creativecommons.org/licenses/by-sa/3.0/

Common name(s): Leathery sea squirt, stalked sea squirt, clubbed tunicate, Pacific rough sea squirt, Asian sea squirt

Global distribution: It is native to the NW waters of Pacific Ocean, including Japan, northern China, Korea, and Siberia.

Ecology: This sessile species is found attached to hard subtidal substrata like buoys, rocks, mussel shells and pilings at depths of 25 m. It is predominantly a littoral zone species although sheltered localities free of floating objects and strong wave action are much preferred by these animals. It is also a hardy species tolerating a wide variations in temperatures (−2–27°C).

Biology: This club-shaped, solitary tunicate species has an oval body which is elongated with a long peduncle which is attached to a substrate. Larger specimens may have a maximum length of 13 cm and smaller specimens which have no distinct peduncle hardly reach 3 cm in length. It has an irregularly grooved skin which is also tough and wrinkled. It occurs in two variations of color depending on size. While the larger animals possess a light brown body with a dark brown peduncle, smaller specimens are invariably yellow-brown.

Nutritional values: This edible species contains nutritionally important amino acids and lipids. The proximate composition, amino acid, and lipid profile of the tissue of this species are shown in Tables 3.19–3.22.

TABLE 3.19 Proximate Composition of *Styela clava* Muscle

Item	%
Moisture	85.6–90.4
Protein	3.2–9.1
Lipid	0.4–1.8
Ash	2.6–2.9
Glycogen	0.4–1.6

Source: Adapted from Park et al. (1991).

TABLE 3.20 Amino Acid Profile of *Styela clava* Tissue

Amino Acid	mg/100 g
Arginine	1–2
Carnosine	4–8
Anserine	1–7
Histidine	1–3
Lysine	6–7
Tryptophan	1
Ornithine	1–2
Beta-alanine	8–17
Phenylalanine	2–3
Tyrosine	1–3
Leucine	3–4
Isoleucine	2–3
Methionine	1–2
Cystine	1–2
Valine	3–5
Alanine	22–46
Glycine	22–112
Proline	12–102
Glutamine	6–20
Glutamic acid	42–60
Asparagine	3–13
Serine	7–10
Threonine	7–12
Hydroxyproline	2–3
Aspartic acid	4–22
Taurine	280–484
Phosphoserine	2–8

Source: Adapted from Park et al. (1991).

TABLE 3.21 Lipid Profile of *Styela clava* Tissue

Item	%
Triglyceride	59.32
Free sterol	23.52
Phosphatidyl choline	49.7
Phosphatidyl ethanolamine	33.0
C20:5 (eicosapentaenoic acid)	17
C22:6 (docosahexaenoic acid)	13.76
C16:0 (palmitic acid)	13.91
C16:1 (palmitoleic acid)	12.52

Source: Adapted from Lee et al. (1995).

Gao et al. (2023) reported that this edible species had total sugar, crude protein, and crude lipid at 1.50%, 8.30%, and 2.1%, respectively.

Styela plicata

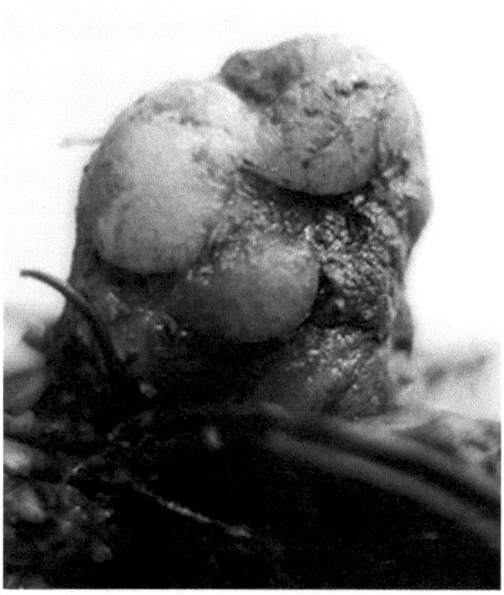

Image credit: Reprinted with permission from Brenda Bowling.

Common name(s): Pleated sea squirt

Global distribution: It has got circum-global distribution.

Ecology: This solitary, benthic species is found attached to hard substrates. It is also reported to occur in a wide range of temperatures (10–30°C) and salinities (22–34‰). Individuals of this species appear either singly or as groups.

Biology: Its oval, upright body is covered by a tough and leathery tunic and is with many rounded warts and pleated grooves. A pair of short siphons is present. Coloration of its body ranges from light tannish white to gray. There are thin red or purple stripes on the insides of the four-lobed siphons; and cross-shaped markings are seen at the tips of the closed siphon. It has an incurrent siphon that intakes water into the pharyngeal basket where food particles like phytoplankton and detritus are filtered out; and the waste is then sent out through the excurrent siphon. It reaches a maximum size of 9 cm.

Nutritional values: This edible species contains nutritionally important amino acids and lipids. The proximate composition, amino acid, and lipid profile of the tissue of this species are shown in Tables 3.22 and 3.23.

TABLE 3.22 Proximate Composition of *Styela plicata* Muscle

Item	%
Moisture	86.5–91.7
Protein	1.7–8.5
Lipid	0.3–1.0
Ash	2.5–2.6
Glycogen	0.4–1.2

Source: Adapted from Park et al. (1991).

TABLE 3.23 Amino Acid Profile of *Styela plicata* Tissue

Amino Acid	mg/100 g
Arginine	9–15
Carnosine	4–8
Anserine	7
Histidine	2
Lysine	5
Tryptophan	1
Ornithine	1
Beta-alanine	36–49
Phenylalanine	2–3
Tyrosine	1–4
Leucine	4
Isoleucine	3–4
Methionine	1–2

TABLE 3.23 *(Continued)*

Amino Acid	mg/100 g
Cystine	2
Valine	3–4
Alanine	22–49
Glycine	25–30
Proline	18–54
Glutamine	8–10
Glutamic acid	29–42
Asparagine	8–10
Serine	8–9
Threonine	7–11
Hydroxyproline	4–6
Aspartic acid	3–8
Taurine	214–533
Phosphoserine	1–8

Source: Adapted from Park et al. (1991).

Bioindicator of pollution: This species has been reported to be a new promising bioindicator of heavy metal pollution (Aydın-Önen, 2016).

Styela rustica

Image not available

Common name(s): Pleated Tunicate, sea potato

Global distribution: Temperate Northern Atlantic, Pacific Ocean, and the Arctic. It is found distributed along the coasts of Mediterranean, Senegal, Brazil, Somalia, Japan, India, North America, the Caribbean, Mexico, Australia, and New Zealand.

Ecology: This sessile, solitary species occurs in mangroves, oyster reefs and rocky areas at a depth range of range 0–432 m. It is also commonly seen on the man-made structures like vessel hulls, marinas, and docks. It is also found attached to the substrate by its posterior body.

Biology: Body of this species is variable in shape, but it is roughly oval. It has a thick, firm, and slightly translucent tunic which is tan or white in color. Tunic forms irregularly rounded, large lumps. While its oral siphon is terminal in position, the atrial siphon is somewhat behind it.

Both the siphons are short and are provided with square-sized apertures with rounded humps on either side. It is a suspension feeder mainly on phytoplankton.

Nutritional values: The edible parts of this species serve as a subsistence food resource called "Oopa" or Upa (as they are called in Russia) to the Native communities in the Bering Strait region. It is added to stews, frozen whole and thinly sliced and consumed with crystalized briny water from inside, or blanched and peeled before consumption (Bugaj, 2022).

Utilization of edible ascidians as a food source in different countries

In Japan, dried *Halocynthia* sp. are added with hot sake in order to make ascidian sake, or they are mixed with hot rice and green tea for making ascidian chazuke.

Korean consumers use *H. roretzi*-flavored whiskey. They also eat raw *H. aurantium* and *H. roretzi* after removing the tunic and dipping the animals into a vinegary and sweet pepper sauce. In Japan, a new product sea squirt called hoya ale is also prepared utilizing.

Tunic-free *Herdmania pallida* is largely utilized in the preparation of ascidian pickle for which salt brine, spices, and vinegar are used. In Korea, pickled *H. roretzi* and *S. clava* form value-added products for marketing. During the pickling and preservation of ascidian, its tiny and sharp calcareous body spicules get dissolved thereby it made suitable for dining. In several Korean recipes, ascidians are mixed with other seafoods, vegetables and meats to provide distinctive flavors and such dishes common at Korean restaurants. Another classic Korean dish called mee-duh-duck chim is prepared by braising or steaming a mixture of *S. plicata* and / or *S. clava* with broth obtained from kombu kelp and dried anchovies; and other condiments. In Korea, *S. clava* powders are utilized in order to increase flavor in soup stocks. A Chilean recipe–stuffed piures is prepared using *P. chilensis* and *P. praeputialis*. Several species of *Microcosmus* are commonly eaten raw in countries around the Mediterranean. In northern Russia, tunic-removed, fresh or frozen *H. aurantium* is chopped and mixed with seafood items such as seaweeds, and whale and seal meat and the entire mixture is eaten raw (Lambert et al., 2016).

CHAPTER 4

PROCESSED ASCIDIANS FOR HUMAN CONSUMPTION

Edible species of ascidians are sold fresh in local markets in many countries. The tunic (test or outer coat) is thick and tough on all the commonly consumed species and is usually removed before preparation (Nanri et al., 1992). However, that the tunic of *M. hartmeyeri* is consumed in some locations in Japan. Bodies removed from the tunic or the whole animals are frozen, canned or dried, and exported for sale worldwide (Lambert et al., 2016). The edible species of ascidians have yielded a number of processed products which are given below.

4.1 ASCIDIAN PICKLES

In the preparation of ascidian pickles, the species, H. pallida, *H. roretzi*, and *S. clava*, have been reported to be more suitable (Tamilselvi et al., 2010; Lambert et al., 2016). The mantle bodies of this species are processed into pickles for human consumption in India. The harvested ascidians after the removal of tunic are cleaned with pure freshwater several times in order to all the epibionts and other wastes from the surface of the organisms. They are then dried in an oven at 80°C. Subsequently, the mantle bodies are removed from the test, followed by cleaning and blanching with a small quantity of turmeric powder, chilly powder, and salt. This preparation is left untouched for about 30 min. In the meantime, fresh garlic, chillies, and gingers are made into small pieces for subsequent processing. Further, fried and powdered fenugreek is also made ready. The quantities of ingredients required in the preparation of ascidian pickles are given in Table 4.1.

Biotechnological Potential of Ascidians: Wacky and Weird Sea Squirts. Ramasamy Santhanam (Author)

TABLE 4.1 Ingredients Required in the Preparation of 500 g Ascidian Pickles

Ingredient	Quantity
Salt	30 g
Chilli powder	20 g
Turmeric powder	2.5 g
Asafoetida	1 g
Garlic	40 g
Mustard powder	5 g
Pepper powder	2 g
Fenugreek	2.5 g
Citric acid	5 g
Sodium benzoate	1 g
Groundnut oil	200 mL
Vinegar	40 mL
Lemon	1
Ginger	Small pieces (cut)
Green chilli	10 g
Curry leaves	3 g

Source: Adapted from Tamilselvi et al. (2010).

Preparation of pickles: Groundnut oil is taken in a pan and is heated for about 180°C. After seeing the boiling oil, mustard powder, and curry leaves are added to the boiling oil and fried till they appear brown. Garlic, green chillies, and lemon pieces are added to the pan one by one with stirring and the pan is kept in sim for about 10 min. Now ginger and asafoetida solution are added and stirred again. Subsequently, vinegar is added. Immediately after the formation of liquid portion, blanched ascidians are added slowly followed by mixing. The gravy will now appear after the addition of 3% blanched solution prepared with citric acid and salt.

After removing the pan from fire, required quantity of preservative, namely, sodium benzoate will be added. Now the pickle is ready and it is packed in sterilized bottles. These pickle bottles are then sealed and stored in refrigerator for future use or sale. Before sealing the bottles, it is better to add some hot groundnut oil to each bottle in order to maintain the quality for about six months. Lambert et al. (2016) reported that the pickling process helps in dissolving the sharp, body spicules thereby the animals are made suitable as an edible item. In Korea, pickled *S. clava* and *H. roretzi* are marketed as commercial products at the price of US$15 for 400 g, and US$32 per 1 kg, respectively (Lambert et al., 2016).

4.2 PIURES

In the preparation of this recipe, piures (*P. chilensis*), the Chilean abalone or ground beef, lapas (*Fissurella*, the large local keyhole limpet), locos (*Concholepas concholepas*), 500 g rice, aji (paprika-dried pepper), cumin, chopped coriander, a large onion, cut lettuce, salt, and white sauce are the ingredients required. As this recipe has three different processes, the amount of piure (ascidians) is first divided into three portions. In the preparation of this recipe, fried stuffing is done first with the onion, cumin, coriander, paprika, and one portion of the selected ascidians. Immediately after this step, another part of the ascidians is washed, and the larger ones are stuffed with the filling as done in the first step. The ascidians are subsequently baked at medium heat on a baking sheet and this is followed by brief broiling. In the final step, rice is prepared with the last portion of ascidians, lepas, and locos or ground beef. While serving each plate is covered with the well-chopped lettuce and is topped with a portion of the rice mixture and stuffed ascidians. White sauce may also be poured over the rice for flavoring (Lambert et al., 2016).

4.3 BOTTLED ASCIDIAN MEAT

Sterilized bottles and irradiated frozen sea squirt meat samples without viscera are used to prepare bottled products with the aim of extending the shelf life of products. Irradiation is done with γ-ray (Co 60) 10 kGy/h of absorbed dose. Sterilization is performed at 121°C for 15 min using an autoclave. The exuviated sea squirt meat is first stored in a cooler (4°C) within 5 min of its purchase. The meat sample is then dehydrated by pressing with a kitchen towel and vacuum-packed (70 and 80 g). The vacuum-packed samples were frozen at −20°C, and these samples were used for the production of bottled meat. In the preparation of this recipe, the meat and the dehydrated meat of sea pineapple, *H. roretzi* are used. Both the meats are mixed with olive oil and sesame oil and are bottled. All the bottled sea squirt meat in oil (BSMO) products are stored at 4°C for 60 days. The oil ratio, meat weight, and added oil volume in both BSMO 1 (meat in oil) and BSMO 2 (dehydrated meat in oil) are shown in Table 4.2.

TABLE 4.2 Oil Ratio, Meat Weight, and Added Oil Volume in h BSMO 1 and BSMO 2

Product	Oil Ratio	Added Olive Oil and Sesame Oil	*Halocyntha roretzi* Meat Weight
BSMO 1	5:5	60 mL	80 g
BSMO 2	7:3	60 mL	80 g

Source: Adapted from Nam-Do et al. (2014).

The nutritional profile (proximate composition) of BSMO 1 and BSMO 2 products is shown in Table 4.3.

TABLE 4.3 Proximate Composition of *Halocyntha roretzi* Meat, Bottled *Halocyntha roretzi* Meat in Oil (BSMO 1), Dehydrated *Halocyntha roretzi* Meat, and Bottled Dehydrated *Halocyntha roretzi* Meat in Oil (BSMO 2)

Item	*Halocyntha roretzi* Meat	BSMO 1	Dehydrated *Halocyntha roretzi* Meat	BSMO 2
Moisture (%)	88.73	75.73	82.94	75.07
Ash (%)	1.37	2.00	2.84	2.13
Carbohydrate (%)	3.02	10.5	4.01	10.44
Protein (%)	6.04	10.04	9.67	10.23
Lipid (%)	0.84	1.73	0.54	2.13

Source: Adapted from Nam-Do et al. (2014).

4.4 FREEZE-DRIED (LYOPHILIZED) ASCIDIANS

In Japan, cultured or wild-harvested *H. aurantium* has been reported to be freeze-dried for the preparation of food supplement tablets or as vacuum-packed and dried strips. Further, the whole ascidian body is also consumed as a snack (Lambert et al., 2016). Further, addition of freeze-dried powder of the ascidians in the food has been reported to enhance the nutritional properties of such food items. Inanami et al. (2001) reported that many quinone derivatives in the ascidians get preserved during freeze–drying process and such natural products possessed the superoxide scavenging activity. Therefore, as a supplement, freeze-dried powder of the ascidians may be of great use in enhancing the antioxidative property of foods.

Freeze drying (lyophilization or cryodesiccation): The main purpose of freeze drying of ascidian-related products is believed to extend their shelf-life besides maintaining the quality. It involves the removal of water from the ascidian product by sublimation. During this process, the ice of frozen ascidian product gets converted to the gaseous state and there is no liquid phase in this process. Further, the rehydrated product is believed to retain much of its original qualities due to the use of low temperature in this processing method. Furthermore, freeze-dried products could be rehydrated or reconstituted easily and much more quickly because this process leaves microscopic pores. Actually four stages, namely, pretreatment, freezing, primary drying, and secondary drying are involved in the entire freeze drying process (McGee, 2015).

4.5 CANNED ASCIDIANS (CANNED "SQUIRT CAN")

The species of sea pineapple, namely, *H. roretzi* is commonly used in the canned ascidian products. The canned "squirt can" using the "squirt" has been reported to be sweet and spicy with gochujang which is a red chili paste. As this paste is spicy and sweet, it is very popular in Korean cooking. The canned ascidian food can be stored at room temperature and it may be used as a snack or side dish.

In the canned ascidian products, the texture of the meat has been reported to be surprisingly soft and crisp. Further, in these products, a complex taste of sea squirts, such as slight bitterness and acidity is present and this taste is relatively easy for everyone to accept. Further, this luxury item seems to go well with sake and white rice (McGee, 2015).

4.6 SAKE-STEAMED HOYA (SEA SQUIRT)

In the preparation of sake-steamed hoya, the sea pineapple, *H. roretzi* (Miyagi Hoya) is used. In the preparation of this product, total cooking time is only around 5 min. This product is eaten after just defrosting. This product with its slightly spicy taste is consumed with the Japanese rice wine (sake) (McGee, 2015).

4.7 HOYA SASHIMI AND HOYA SASHIMI WITH HALF SHELL

4.7.1 HOYA SASHIMI

In the preparation of this product, freshly picked Miyagi Hoya (sea pineapple) is frozen immediately after its processing and peeling. In this way, this product retains the same sweetness and taste resembling Miyagi Hoya. It may be consumed as sashimi which is a Japanese delicacy made of sliced meat or raw fish. This product is eaten after just defrosting (McGee, 2015).

4.7.2 HOYA SASHIMI WITH HALF SHELL

This product is considered to be the best for one who eats at home without using the kitchen. In order to display the characteristics of Hoya, a bowl of shell is put on it. It is also reported that even a small portion of ponzu (i.e., Japanese sauce made of citrus juice and soy sauce) mixed with Hoya juice would increase the taste considerably. This product is eaten after just defrosting (McGee, 2015).

4.8 HOYAN PRA (JAPANESE NAM PLA)

This ascidian extract is prepared in the same manner as fish sauce and for its production, it requires only three to five months. It is also interesting to know that even a few drops of this extract may create the entire flavor of the dish like Ramen which is fried rice and pasta. In the production of this Hoyan Pra, no chemical seasoning added (McGee, 2015).

4.9 ASCIDIAN ALE (HOYA ALE): BLUE BEER

In the production of this ascidian beer, malt and hops are added to a broth filled with sea pineapples (*H. roretzi*), husks, and others. The produced beer is a red amber ale which is said to have a smoky aroma and a clean, refreshing finish. This beer has been reported to hit the market in Japan (McGee, 2015).

CHAPTER 5

ASCIDIANS AS A POTENTIAL SOURCE OF AQUACULTURE, AQUARICULTURE, AND LIVESTOCK FEEDS

The aquaculture industry assumes greater importance in mitigating the nutritional problems of the ever-increasing human population. In order to ensure sustainable development of this industry, there is an urgent need in searching alternative protein sources for the presently and commonly used fishmeal and plant-derived ingredients. Further, a thorough knowledge is essential on the nutritional requirements and production of fish feed, especially in the development and sustainability of aquaculture industry. In this regard, several species of ascidians with their enhanced protein and/or lipid levels have been reported to serve as a potential ingredient in animal feeds for feeding fish and shrimps in aquaculture systems, ornamental fish in aquariums, and livestock such as pigs (https://patents.google.com/patent/WO2013088177A1/en). Further, the ascidians could serve as a sustainable and alternate source of quality protein as a substitute for fishmeal, especially in the culture of Atlantic salmon (Gao et al., 2023). Several species of washed or dried whole ascidians or their extracts have been reported to serve as ingredients in the production of aqua and livestock feeds. On the other hand, the tunic with cellulose of very few species like *H. roretzi* is utilized in fish and other livestock feeds (https://patents.google.com/patent/WO2013088177A1/en). The biology and ecology of species of ascidians serving as ingredients in the aquafeeds are given below.

Biotechnological Potential of Ascidians: Wacky and Weird Sea Squirts. Ramasamy Santhanam (Author)

Ascidia gemmata

Ascidia sp.

Image credit: Gronk. https://creativecommons.org/licenses/by-sa/3.0/

Common name(s): Not designated

Global distribution: Tropical Indo-West Pacific

Ecology: This sessile species is commonly found on stones of intertidal areas and mangrove roots of marine lakes at a depth range of 1–12 m.

Biology: This translucent jelly-like animal grows up to 3.5 cm in total length and is found fixed to the substrata by the left side. Body is orange in coloration, and it may have some epibionts such as algae and bryozoans on its body surface. Tunic is cartilaginous of 0.6–1.8 mm thick and is slightly wrinkled and translucent. There are small conical projections which are irregularly distributed on the surface of the animal, each one corresponding to the end of blood vessels. Both siphons are 0.3–0.5 cm long. While the oral siphon has eight smooth lobes, the atrial siphon has six smooth lobes. There are 58–69 oral tentacles which project from a thick muscular ring.

Animal feed values: In washed dried for or as extracts, this species has the potential use of serving as an ingredient in the aquafeed preparation (Gao et al., 2023).

Ascidia mentula

Image credit: Matthieu Sontag. https://creativecommons.org/licenses/by-sa/3.0

Common name(s): Pink sea squirt

Global distribution: The Mediterranean Sea, Black Sea and Northeast Atlantic Ocean.

Ecology: It is commonly found on rocky substrata at depths down to 200 m. It is known to prefer gullies and crevices and also adheres to shells, stones, and seaweed, *Laminaria* holdfasts.

Biology: It is a solitary ascidian and its size ranges from about 5 to 30 cm in length. Test is generally pink and is smooth, thick and cartilaginous. Its ovoid body is found fixed to the substrata by its left side. Oral siphon is broad, terminal and eight-lobed; and it may have eight white markings. Atrial siphon is one-half of body length from the oral siphon. It feeds on plankton which is filtered from the seawater using its mucous net. The bivalve mollusk, *Modiolarca tumida,* is found living within the tunic of this animal. Further, the pea crab, *Pinnotheres pinnotheres*, and copepod, *Notopterophorus papilio*, have also been reported to occupy its body cavity. This species has been reported to live up to seven years.

Animal feed values: In washed dried for or as extracts, this species has the potential use of serving as an ingredient in the aquafeed preparation (Gao et al., 2023).

Ascidia virginea

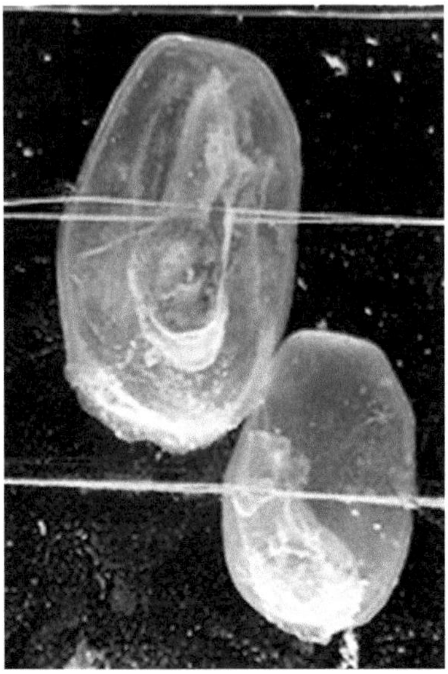

Image credit: Daderot. Public domain

Common name(s): Not designated

Global distribution: Temperate and subtropical Northeast Atlantic and the Mediterranean. It is also widespread around the British Isles.

Ecology: This sessile specie is commonly found at a depth range of 41–44 m in the rocky substrata of semi-exposed silty or sheltered areas.

Biology: This rectangular-shaped, solitary ascidian is found attached to the substrata by a small area of its base. Test is firm and thick, and it is very clean with milky white or pale pink coloration. Oral siphon is terminal and its atrial siphon is slightly down the side of its body. It has a size of 6 cm tall × 3 cm wide.

Animal feed values: In washed dried for or as extracts, this species has the potential use of serving as an ingredient in the aquafeed preparation (Gao et al., 2023).

***Ascidia* sp.**

Animal feed values: The fatty acid profile (vide Table 5.1) of this unidentified species has been reported to be of great use in the development of aquafeeds (Gao et al., 2023)

TABLE 5.1 Fatty Acid Profile of *Ascidia* sp. Usable in Aquafeed Formulation

Fatty Acid	Percentage
C14:0	3.76
C15:0	1.18
C16:1	6.35
C16:0	20.94
C17:0	1.18
C18:4	1.41
C18:2	0.71
C18:1	25.41
C18:0	12.71
C20:5	6.82
C20:4	2.35
C20:2	1.65
C20:1	7.76
C20:0	0.47
C22:6	2.59
C22:1	2.12
C22:0	2.59
Saturated fatty acids	42.83
Unsaturated fatty acids	57.17
Monounsaturated fatty acids	41.64
Polyunsaturated fatty acids	15.53

Source: Adapted from Gao et al. (2023).

Ascidiella aspersa

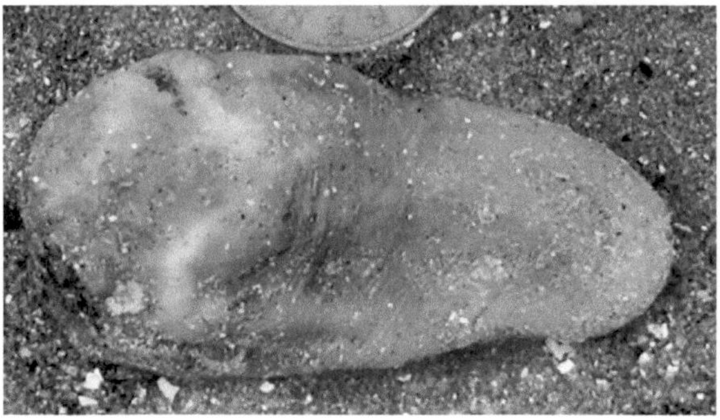

Image credit: Notafly. https://creativecommons.org/licenses/by-sa/3.0/

Common name(s): European sea squirt

Global distribution: The Northeast Atlantic from the Mediterranean Sea to Norway is the origin of this species.

Ecology: This solitary sea squirt is found attached to the substrata by its left side. It is found distributed in dense groups on hard substrata including rocks at depths up to 90 m. At depths of 0–90 m, it is found distributed on hard substrata like rocks in the form of dense groups of free individuals.

Biology: This oval bodied species grows up to 13 cm in length. Its branchial or oral siphon is conical and is situated at its top. Six to eight lobes are seen. Atrial siphon is situated at the side of its body and is with six lobes. Its body covering, namely, tunic is transparent and is grayish to brown in coloration. When expanded, about 40 tentacles can be seen on the inner surface of its branchial wall. Both the branchial and atrial openings possess faintly colored ridges on their rims. This species is distinguished from its closely resembled species, namely, *C. intestinalis* by the lack of yellow markings around its siphons.

Animal feed values: In washed dried for or as extracts, this species has the potential use of serving as an ingredient in the aquafeed preparation (Gao et al., 2023).

Botryllus schlosseri

Image credit: Massimiliano DE MARTINO. https://creativecommons.org/licenses/by-sa/3.0/

Common name(s): Golden star tunicate, star tunicate, or star ascidian

Global distribution: It is a native species of Northeast Atlantic Ocean, North Sea, and the Mediterranean Sea.

Ecology: It commonly occurs on the exposed rock surfaces of the lower shore, on seaweeds, and kelp stipes.

Biology: Colony of this species may reach a size of 5 cm in length and individual zooids of may grow up to 3 mm in size. Zooids arise from a center as arms of a star. Each cluster of the colony may have 5–8 zooids. Coloration of the colonies of this species may be orange, blue, or gray. An independent colony derived from the main body of an existing colony is termed as a subclone.

Animal feed values: In washed dried for or as extracts, this species has the potential use of serving as an ingredient in the aquafeed preparation (Gao et al., 2023).

Ciona intestinalis

Common name(s): Vase tunicate

Global distribution: This cosmopolitan species has got global distribution.

Ecology: It is found as dense aggregates on submerged or floating substrata, particularly artificial substrata like floats, boat hulls, and pilings, in the lower intertidal to subtidal shores. It is also often grown with other fouling organisms. It is believed to occupy fresh and new areas mainly by hull fouling.

Biology: Body of this solitary ascidian is cylindrical, soft, and gelatinous and is largely resembling an intestinal mass sprouting from a rocky substratum. The body of the animal is fixed to the substratum by its permanent base situated at the posterior end. Top of the animal possesses two openings, namely, the buccal and atrial siphons. Its tunic is very soft. Maximum length of this species is about 20 cm.

Animal feed values: The meal of this species is derived from its inner tissues which constitute 55% of total animal weight. This prepared meal has been reported to contain 41–45% marine proteins and 3–3.51% marine lipids in which 48% constitutes omega-3 fatty acids (Kousoulaki et al., 2022). Based on these findings, this species has been reported to be the commonest species used in the aquafeed preparation owing to its fatty acid and amino acid composition as shown in Tables 5.2–5.5. Its internal organs also possess the highest protein content (69.3%) and the lowest cellulose (5.6%).

Further, in this species, the ratios of EEAs (essential amino acids) to the TAAs (total amino acids) of all the internal organs were in the order of 3250%. Furthermore, the essential amino acid index which is representing the similarity of the ascidian species' amino acid composition to that of egg protein, was also found to be high. These findings suggest that the protein of this species is of very high quality and is more suitable for aquafeed formulation and preparation (Gao et al., 2023). As per available reports, *Cliona* meal is a novel marine feed ingredient as it becomes a 100% organic product in a zero-emission circular system after its harvest. Further, this meal is an excellent ingredient in feed production, as it contains healthy marine fats and minerals needed for the culturable finfish and shellfish fish as well as for aquarium fish. Furthermore, this ciona meal can replace fish meal without losing the most important substances for the favorable growth of the shrimp and fish (https://marinefeed.com/products-ciona-meal/)

TABLE 5.2 Fatty Acid Profile of *Ciona intestinalis* Usable in Aquafeed Formulation

Fatty Acid	Percentage
C14:0	2.08
C15:0	0.83
C16:1	8.33
C16:0	11.67
C17:0	1.25
C18:4	0.83
C18:2	1.67
C18:1	14.58
C18:0	6.25
C20:5	24.58
C20:4	0.42
C20:2	0.83
C20:1	10.83
C20:0	1.67
C22:6	12.92
C22:1	1.33
C22:0	0.33

TABLE 5.2 *(Continued)*

Fatty Acid	Percentage
Saturated fatty acids	24.08
Unsaturated fatty acids	76.32
Monounsaturated fatty acids	35.07
Polyunsaturated fatty acids	41.25

Source: Adapted from Gao et al. (2023).

TABLE 5.3 Essential Amino Acid Contents in the Whole Body of *Ciona intestinalis*

Essential Amino Acid	% in Whole Body
Arg	2.13
His	0.9
Ile	1.28
Leu	2.54
Lys	1.84
Met	0.54
Phe	1.13
Thr	2.17
Val	1.7
Total	14.23

Source: Adapted from https://marinefeed.com/products-ciona-meal/

TABLE 5.4 Nonessential Amino Acid Contents in the Whole Body of *Ciona intestinalis*

Nonessential Amino Acid	% in Whole Body
Ala	1.81
Asp	4.18
Cys	1.93
Glu	5.27
Gly	2.31
Pro	1.77
Ser	2.18
Tyr	1.56
Total	21.01

Source: Adapted from https://marinefeed.com/products-ciona-meal/

TABLE 5.5 Comparison of Amino Acid Contents of *Ciona intestinalis* and Egg Albumin

Amino Acid	*Ciona intestinalis* (g/100 g of Protein)	Egg Albumin (g/100 g of Protein)
Ala	4.96	5.7
Arg	7.00	5.9
Asp	11.11	9.2
Cys	3.28	–
Glu	16.43	15.7
Gly	7.35	3.2
His	2.26	2.41
Ile	3.32	7.1
Leu	7.31	9.9
Lys	6.54	6.4
Met	2.10	5.4
Phe	3.34	7.5
Pro	4.76	3.8
Ser	5.79	8.5
Thr	5.13	4.00
Tyr	3.82	3.75
Val	4.01	8.8
Total	98.53	107

Source: Adapted from https://marinefeed.com/products-ciona-meal/

***Ciona* meal as Atlantic salmon and livestock feed:** The Norwegian research institute Nofima conducted feeding tests with Atlantic salmon for three months with feed where 50% of the fish meal was replaced with *Ciona* meal. Based on the fruitful results, *Ciona* meal was verified as a fish feed ingredient by the aforesaid institute. Further, *Ciona* meal was also tested and verified as feed ingredient for pigs, chickens, and cats by the Swedish University of Agricultural Sciences during 2018 and 2019. *Ciona* meal is also now accepted as an EU-registered feed ingredient (https://feedmaterialsregister.eu/register). Gao et al. (2023) reported that the Atlantic salmon aquaculture exhibited improved growth and fatty acid profiles when half of the dietary fish meal protein was substituted with the ascidian, ciona meal at a concentration of 17%. Further, it was stated that the ciona meal could reduce pollutants such as heavy metals (e.g., arsenic and vanadium) and toxins (e.g., toxaphene) in the whole body of salmon.

Clavelina lepadiformis

Image credit: Pino Bucca. https://creativecommons.org/licenses/by-sa/3.0/

Common name(s): Light-bulb sea squirt

Global distribution: This species is native to the Northeast Atlantic Ocean. It commonly occurs in Great Britain and Ireland; and all along European coasts south of the Mediterranean Sea.

Ecology: It is shallow water, epibenthic, colonial ascidian found in the subtidal and low intertidal zones.

Biology: Colony of this species is large, and it may consist of hundreds of zooids. Fully grown zooids of this species are 5–6 cm long and are resembling light bulbs without colorful bands around the dorsal lamina and oral siphon and dorsal lamina. All the elongated zooids are partially or wholly embedded in tunic and are united by basal and stolons. Each zooid is with an oral and atrial siphon, which helps open the colony. Body is divided into two regions, namely, a thorax and abdomen. While the thorax contains the pharynx, the abdomen which is below thorax contains the stomach, intestine, heart, testis, and ovaries.

Animal feed values: The crude cellulose and chemically pure cellulose yielded by this species have been reported to be 96% and 54%, respectively. Further, its protein in the treated residue was found to be 6%. All these factors make this species as a very suitable ingredient for aquafeed preparation (Hassanzadeh, 2011).

Corella parallelogramma

Corella sp.

Image credit: Randal. https://creativecommons.org/publicdomain/zero/1.0

Common name(s): Gas-mantle ascidian

Global distribution: North Sea, British Isles, and from Norway to the Mediterranean.

Ecology: This species grows on stones, rocks, shells, and seaweed, *Laminaria* in areas of strong currents. It attaches itself to substratum by a small part of its base. It is also known to prefer clear water.

Biology: This rectangular and laterally compressed, solitary ascidian has a thin, transparent test (tunic). Individuals of this species may be 30–50 mm in length and 20–30 mm in width. Gut and branchial sac (i.e., gill) are quite visible. Branchial sac is well marked with red or yellow pigments which are occurring in a criss-cross pattern. Atrial siphon is usually short but rarely it may be very long in certain individuals. Oral siphons possess ocelli (red eyespots). Gut is situated on the right side of the gill. Intestine, a gray tube, passes along the base.

Animal feed values: In washed dried for or as extracts, this species has the potential use of serving as an ingredient in the aquafeed preparation (Gao et al., 2023).

Cystodytes violatinctus

Common name(s): Purple colonial ascidian

Global distribution: Tropical Indo-West Pacific.

Ecology: This sessile and benthic species occurs at a depth range of 7–40 m.

Biology: No report

Animal feed values: In washed, dried or as extracts, this species has the potential use of serving as an ingredient in the aquafeed preparation. The fatty acid profile of this species as reported by Gao et al. (2023) is given in Table 5.6.

TABLE 5.6 Fatty Acid Profile of *Cystodytes violatinctus* Usable in Aquafeed Formulation

Fatty Acid	Percentage
C12:0	1.20
C14:0	4.50
C15:0	1.50
C16:1	1.26
C16:0	21.20
C17:0	4.80
C18:1	20.00
C18:0	24.80
C19:0	3.60
C20:0	3.10
C21:0	2.30
C22:0	0.40

Source: Adapted from Gao et al. (2023).

Halocynthia aurantium

Image credit: Erin McKittrick. https://creativecommons.org/licenses/by/4.0

Common name(s): Pacific sea peach

Global distribution: It is a polar and subtropical species and is found distributed in the Northern Pacific Ocean. It is also found to be a very common species of the Bering Sea.

Ecology: This sessile and subtidal species is found on rocks, pilings, floats, and kelp beds and in tide-pools at depths ranging from 0 to 100 m. Its body is attached directly to the substrate.

Biology: It is typically a barrel-shaped organism and is attaining a maximum height of 18 cm. Its tunic is smooth or wrinkled. Body coloration is usually peachy, red, or orange; and unique and beautiful with iridescent quality. Both oral and atrial siphons are seen at the top. The sea peach is largely preyed upon by the sea stars and crabs.

Animal feed values: In washed, dried or as extracts, this species has the potential use of serving as an ingredient in the aquafeed preparation. The fatty acid profile of this species as reported by Gao et al. (2023) is given in Table 5.7.

TABLE 5.7 Fatty Acid Profile of *Halocynthia aurantium* Usable in Aquafeed Formulation.

Fatty Acid	Percentage
C14:0	4.90
C15:0	4.60
C16:0	18.10
C17:0	6.50
C18:0	18.60
C19:0	2.30
C20:0	5.00
C22:0	25.80
C23:0	4.50
C24:0	2.10

Source: Adapted from Gao et al. (2023).

Halocynthia roretzi

Vide Chapter 3 for details.

Animal feed values: In washed, dried, or as extracts, this species has the potential use of serving as an ingredient in the aquafeed preparation. The fatty acid profile of this species as reported by Gao et al. (2023) is given in Table 5.8.

TABLE 5.8 Fatty Acid Profile of *Halocynthia roretzi* Usable in Aquafeed Formulation

Fatty Acid	Percentage
C14:0	4.10
C15:0	6.10
C16:1	2.10
C16:0	25.40
C17:0	7.20
C18:0	16.80
C20:0	2.50
C22:1	4.10

Source: Adapted from Gao et al. (2023).

Polyclinum aurantium

Image credit: Matthieu Sontag. https://creativecommons.org/licenses/by-sa/3.0/

Common name(s): Not designated

Global distribution: This temperate and subtropical species is native to the northeastern Atlantic Ocean and Mediterranean Sea. It is also common in European Atlantic and Western Africa.

Ecology: This shallow water species occurs on rocks, stones and other substrates like algal beds at a depth range of 0–100 m. It is commonly found in wave exposed gullies and in the upper infralittoral zone.

Biology: This round and pear-shaped colonial species has been reported to form flat-topped or globular mounds which consist of several zooids. All these zooids are seen immersed in a common tunic (test). Colony is generally 1–1.5 cm high and 2–5 cm wide. Each zooid of the colony has a buccal siphon which is with six lobes. Stomach wall of the colony is smooth and its intestinal loop is twisted. Colony has a few common cloacal siphons, each of which has a long tongue-like projection. Individual zooids are up to 6 mm long, and coloration of the colony is yellowish-gray or yellowish-brown and body of the coly is often found coated with sand.

Animal feed values: In washed, dried or as extracts, this species has the potential use of serving as an ingredient in the aquafeed preparation (Gao et al., 2023).

Pyura squamulosa *(=Cynthia squamulata)*

Pyura sp.

Image credit: Richard Ling. https://creativecommons.org/licenses/by-sa/2.0/

Common name(s): Red sea squirt, scaly sea squirt

Global distribution: This subtropical species is found distributed throughout the British Isles, north-east of the Atlantic, European waters, and the Mediterranean.

Ecology: This sessile species has a depth range of 41–44 m.

Biology: It is a solitary sea squirt and its oral siphon is tetralobular with branched tentacles. It grows up to a size of 2.5 cm in length. It has been reported to harbor the copepod, *Bonnierilla similis.*

Animal feed values: In washed, dried or as extracts, this species has the potential use of serving as an ingredient in the aquafeed preparation. The fatty acid profile of this species as reported by Gao et al. (2023) is given in Table 5.9.

TABLE 5.9 Fatty Acid Profile of *Pyura squamulosa* Usable in Aquafeed Formulation

Fatty Acid	Percentage
C12:0	2.36
C14:0	3.71
C15:0	4.69
C16:1	6.14
C16:0	5.83
C17:0	6.47

TABLE 5.9 *(Continued)*

Fatty Acid	Percentage
C18:4	12.14
C18:3	9.76
C18:2	8.97
C18:1	8.85
C18:0	8.53
C20:5	20.74
C20:5	17.35
C22:6	40.13

Source: Adapted from Gao et al. (2023).

Styela plicata

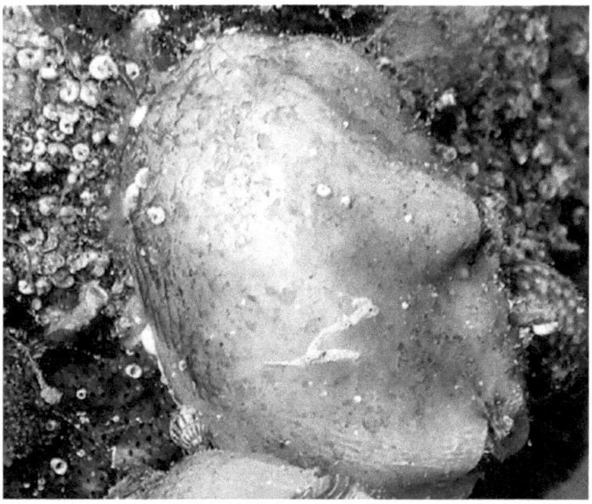

Image credit: Frédéric André. https://creativecommons.org/licenses/by-sa/4.0/

Vide Chapter 3 for details.

Animal feed values: In this species, the ratios of the essential amino acids to the total amino acids of all the internal organs were in the order of 32–50%. Further, the essential amino acid index which is representing the similarity of the ascidian species' amino acid profile to that of egg protein was also found to be high. The very high-quality protein and lipid profile of this species make it highly suitable for aquafeed formulation and preparation (Gao et al., 2023).

The fatty acid profile of this species is given in Table 5.10.

TABLE 5.10 Fatty Acid Profile of *Styela plicata* Usable in Aquafeed Formulation

Fatty Acid	Percentage
C14:0	0.68
C15:0	2.04
C16:1	4.08
C16:0	15.31
C17:0	7.48
C18:4	0.34
C18:2	1.36
C18:1	15.65
C18:0	11.22
C20:5	25.85
C20:4	0.34
C20:2	1.7
C20:1	3.06
C20:0	0.34
C22:6	5.78
C22:1	3.4
C22:0	1.36
Saturated fatty acid	38.43
Unsaturated fatty acids	61.56
Monounsaturated fatty acids	26.19
Polyunsaturated fatty acids	35.37

Source: Adapted from Gao et al. (2023).

5.1 CHEMICAL CONSTITUENTS OF COMMON SPECIES OF ASCIDIANS USABLE IN AQUAFEED PRODUCTION

The ascidian body is made of two main parts, namely, the outer shell (tunic or test) and the internal organs. The tunic contains about 60% cellulose and nitrogen-containing organic ingredients which comprise about 27% by dry weight. On the other hand, the maximum protein and lipid contents in the body tissues of the different species of ascidians were found to be 41% and 3% of dry weight, respectively. It is worth mentioning here that the internal organs of several species of ascidians are abundant with good proteins which are indicated by high contents (more than 90%) of essential and delicious amino acids. Further, the amino acid composition of many species of ascidians is found to be similar to that of egg albumin, and therefore these ascidians are of great value to be considered as an active ingredient of marine aquafeeds (Gao et al., 2023). Marques et al. (2022) reported that the substitution of fish meal with essential amino acid-rich ascidian meal in aquaculture feeds has already commenced in the farming practices of several countries. In regard to the fatty acid profile of the ascidians, it seems to be more or less similar to fish oil. Further, the PUFAs, namely, EPA (20:53), and docosahexaenoic acid (DHA, 22:63) contents in the ascidians are also high; and the levels of DHA, EPA, and arachidonic acid, of the ascidian meals were found to be 22.8%, 17.2%, and 3.5% of total fatty acids, respectively (Gao et al., 2023). Keeping these in consideration, the ascidians meal may be a good and viable alternative to fish meal and fish oil in the formulation of aquafeeds. The chemical constituents of the most potential ascidian species usable in aquafeed preparation are given below.

Halocynthia roretzi: In this species, the total lipid content is 2.0%. Of the total lipids, 17.2% are glycolipids; 36.6% are neutral lipid and 46.2% are phospholipids. Among the neutral lipids, the free sterols constitute 25.8%, and triglycerides, 49.0%. Other lipids include 5.2% esterified sterols and hydrocarbons; 6% monoglycerides, 9.4% diglycerides, and 4.6% free fatty acids. The phospholipids comprise 48.6% phosphatidylcholine, 32.4%, phosphatidylethanolamine, 5.7% phosphatidylserine, 9.8% phosphatidylinositol, and 3.5% unknown compounds. The primary fatty acids available in the total lipids are 2% and 8.5%, 1.3% eicosapentaenoic acid, 16.3% docosahexaenoic acid, 13.8% palmitic acid, 8.5% oleic acid (Gao et al., 2023). Marques et al. (2022) reported that the tunic of this ascidian species could partially or wholly substitute the common fish meal in aquaculture feeds for farming the molluskan species *H. discus* for the better growth of the latter.

Halocynthia auantium: The tunic of this species consisted mainly of palmitic acids (21.73%), oleic acid (6.78%), stearic acid (33.13%), lauric acid (2.72%), dihomo-linolenic acid (4.09%), DHA (3.38%), and EPA (3.88%). Amazingly, both the outer membrane and inner membrane of its tunic displayed significant levels of -3 PUFAs than -6 PUFAs. Further, the concentrations of EPA were found to be higher than that of

DHA. Another study reported that the lipids of the tunic of this species comprised of different types of valuable lipids, such as glycolipids, phospholipids, and neutral lipids. These lipids are mainly composed of SAFA (16:0 and 18:0), MUFA (16:17 and 18:19), and PUFAs (18:26 and 20:53) (Gao et al., 2023).

***Ciona intestinalis*:** As its inner body tissue has a high protein content (up to 47%), followed by ash (32%) and carbohydrates (4%), it is ideal for aquafish feed formulations. The use of this *Ciona* meal for fish feed is considered to be more sustainable and environmentally friendly compared to the traditional approach of feeding fish such as salmon with industrially caught fish. Thus, cultivation of this species ascidians is becoming popular and several ascidian farms have now been established. For example, a Swedish company called marine feed has developed a *Ciona* cultivation technology that has achieved a harvesting capacity of 10 t/day. In 2017, 180 t of wet ascidian biomass was harvested. Moreover, it has been estimated that cultivation of *Ciona* can provide around 250 t of dry weight per hectare of the ocean surface per year, which is considerably more than any of the land crops (Hrůzová et al., 2020).

***Botryllus schlossen*:** It has been reported to contain 18 sterols and the major sterols are stanols. Though its phospholipid profile is complicated, phosphatidylethanolamine, and phosphatidylcholine are the major representatives (Gao et al., 2023).

5.2 TRACE ELEMENTS IN ASCIDIANS USABLE IN AQUAFEED PRODUCTION

Several species of ascidians including *H. aurantium* have been reported to accumulate both essential trace elements necessary for their physiological functions and heavy metals with toxic properties (Gao et al., 2023). Apart from protein and lipids, several species of ascidians including *C. intestinalis* have a unique capacity of accumulating heavy metal ions from the environment. Among these metals, minerals, particularly, Mg, Zn, and V, are believed to be beneficial and are fit for exploitation. It is also well known that ascidians concentrate V (vanadium) from the seawater through vanadocytes. Further, the phlebobranchian ascidian is known to contain more of vanadium than those of the stolidobranchian ascidians. It is reported that the blood cells of *Ascidia gemmata* exhibited the highest quantities of vanadium (i.e., 350 mM) which is about 100 times the vanadium concentration of seawater. Similarly, *H. aurantium* has been reported to contain several trace elements such as Cu, Pb, Fe, Zn, Mn, Ni, and Cd in its tissues including the tunic, muscular sac, digestive gland, gonads, and stomach,. The highest levels of the biologically significant elements were, however, noticed in the tunic (Mn), stomach (Zn and Cu), digestive gland (Zn), and gonads (Cu). Further, the manganese levels in the tunic of this species were also found to be higher than in other organs. Based on these findings, this species is considered to be an ideal ascidan biomass for use in aquafeed applications.

5.3 AQUAFEEDS: FORMULATION AND CHARACTERISTICS

Basic composition of aquafeeds: Prepared, complete aquafeeds may have moisture (<10%), carbohydrate (15–20%), protein (18–50%), lipid (10–25%), ash (<8.5%), phosphorus (<1.5%), vitamins, and minerals in traces (Craig and Helfrich, 2017).

Fatty acids in ascidians and fish feeds: Marques et al. (2021) reported on the general nature of the fatty acid content of ascidians and fish feeds as shown in Table 5.11.

5.4 FORMULATION AND PREPARATION OF EXPERIMENTAL AQUAFEEDS USING ASCIDIAN MEAL

Four experimental feeds were formulated and prepared using the body tissues of the ascidian, *C. intestinalis*. In these feeds, the levels of ascidian meal, soy protein meal, and fish meal were only varying and the levels of other ingredients were kept same. According to the research findings, 64% of the fish meal can be replaced by the ascidian meal without compromising the physical feed quality. Further, the ascidian meal was also found to improve the pellet expansion and fat absorption capacity of the extruded feed and this is largely due to the viscosity behavior of the chosen species of ascidian, namely, *C. intestinalis* (Samuelsen et al., 2022). The concentrations of the different constituents and the proximate composition of the different feeds so prepared are shown in Tables 5.12 and 5.13.

TABLE 5.11 Fatty Acids in Ascidians and Fish Feeds

Fatty Acid	Molecular Formula	Ascidians (%)	Fish Feeds (%)
Arachidonic acid	20:4n-6	3.11	0.47
Docosahexaenoic acid	22:6n-3	11.85	4.59
Eicosapentaenoic acid	20:5n-3	20.44	2.13
Hexadecenoic fatty acid	16:1n-9	5.37	3.62
Linoleic acid	18:2n-6	2.26	16.86
α-Linolenic acid	18:3n-3	2.38	2.85
Myristic acid	14.0	1.45	1.53
Omega-7 fatty acid	18:1n-7+n-9	11.98	35.97
Palmitic acid	16.0	12.56	17.25
Stearic acid	18.0	5.89	6.51
Stearidonic acid	18:4n-3	3.61	0.62

Source: Adapted from Marques et al. (2021).

TABLE 5.12 Formulation of Aquafeeds Using the Ascidian, *Ciona intestinalis*

Feed Ingredient	Feed 1 (g/kg DM)	Feed 2 (g/kg DM)	Feed 3 (g/kg DM)	Feed 4 (g/kg DM)
Ascidian	172	345	172	115
Wheat flour	138	138	138	138
Wheat gluten	215	215	215	215
Soy protein concentrate	0	0	172	115
Fish meal	172	0	0	115
Corn gluten meal	69	69	69	69
Rapeseed oil	56	56	56	56
Faba (bean)	110	110	110	110
Vit. mixture	7	7	7	7
Min. mixture	7	7	7	7
DL-Methionine	3	3	3	3
L-Lysine	14	14	14	14
L-Histidine	3	3	3	3
Monosodium phosphate	34	34	34	34

Source: Adapted from Samuelsen et al. (2022).

TABLE 5.13 Proximate Composition[a] of Feeds 1–3 Prepared Using the Ascidian, *Ciona intestinalis*

Item	Feed 1 (g/kg DM)	Feed 2 (g/kg DM)	Feed 3 (g/kg DM)	Feed 34 (g/kg DM)
Total ash	148	186	137	128
Carbohydrate	256	298	298	270
Protein	504	433	485	515
Lipid	99	85	81	91

Notes: [a]Calculated at total ash level of 98 g/kg, carbohydrate level of 100 g/kg, protein level of 374 g/kg, and lipid level of 338 g/kg.

Source: Adapted from Samuelsen et al. (2022).

5.5 ROLE OF TUNICS CAROTENOID EXTRACT OF THE SEA PINEAPPLE, *HALOCYNTHIA RORETZI* ON THE GROWTH AND MUSCLE COLOR OF RAINBOW TROUT IN AQUACULTURE SYSTEMS

Preparation of carotenoid extract from *Halocynthia roretzi:* About 2 cm *Halocynthia roretzi* tunics were chopped and soaked in acetone (10× v/w) overnight. Then the acetone was filtered and evaporated using vacuum evaporator. Subsequently, the extract was evaporated until dryness. The resulting carotenoid extract was stored in an amber bottle, sealed under nitrogen gas and kept at −20°C before use. The above tunics had 302.6 mg/kg carotenoids and 19.0% of lipid. The fatty acid profile of tunics so determined is given in Table 5.14.

Growth and muscle color of rainbow trout in aquaculture systems: A total of three experimental feeds were prepared with the following ingredients (Table 5.15).

The feeding experiments were conducted on the rainbow trout (*Oncorhynchus mykiss* with the aforesaid three feeds for a period of 120 days. And after the experimental period, the growth and muscle color of the fish were investigated. Among all the feeds tested, the feed containing carotenoid extract of the sea pineapple showed amazing results. The result showed that Feed 3 (ascidian feed) had the highest final weight (1119.2 g) compared to those of Feed 1 (881.0 g) and Feed 2 (1068.2 g). Further, the specific growth rate (SGR) of Feed 3 (1.0%/day) was significantly higher than Feed 1 (0.7%/day) and Feed 2 (0.7%/day). Furthermore, the feed conversion ratios (FCR) were found to be 1.5, 1.4, and 1.2 for Feed

1, 2, and 3, respectively. The muscle color was also found to be positively affected by the treatments as the muscle color of the experimental fish fed with Feed 2 and Feed 3 were significantly different from Feed 1. The initial muscle color score was 1.7 and the final scores were found to be 3.4, 5.6, and 5.7 for Feed 1, 2, and 3, respectively. Further, the muscle carotenoids content of Feed 3 (8.5 mg/kg) was significantly higher than those of Feed 2 (6.9 mg/kg) and Feed 1 (6.1 mg/kg) (Rohmaha et al., 2016).

TABLE 5.14 Fatty Acid Profile of *Halocynthia roretzi* Tunics Carotenoid Extract

Fatty Acid	Percentage
16:1n7	8.2
18:1n9	5.6
18:1n7	6.8
20:1n7	1.8
22:1n13	1.1
22:1n9	1.0
22:1n7	1.0
Total monounsaturated	27.8
16:2n6	0.6
18:2n6	2.2
18:3n3	1.5
18:3n1	0.2
18:4n3	1.0
20:3n6	0.1
20:3n4	0.4
20:3n3	1.0
20:4n6	4.5
20:5n3	7.9
22:5n6	1.0
22:6n3	5.8
Total polyunsaturated	34.3
12:0	0.5
14:0	4.7
16:0	17.6
18:0	6.8
19:0	0.2
20:0	0.4
22.0	0.52
Total unstaturated	37.9

Source: Adapted from Rohmaha et al. (2016).

TABLE 5.15 Proximate Composition/Ingredients of Experimental Feeds

Ingredients	Exptl. Feed 1	Exptl. Feed 2	Exptl. Feed 3
Moisture	7%	7%	7%
Ash	8.5%	8.5%	8.5%
Protein	48.1%	48.1%	48.1%
Lipid	23.9%	23.9%	23.9%
Astaxanthin	39.5 mg/kg	39.5 mg/kg	39.5 mg/kg
Synthetic canthaxanthin (Carophyll Red® 10%)	–	10 mg/kg	–
Carotenoid extract of tunic	–	–	10 mg/kg

Source: Adapted from Rohmaha et al. (2016).

Application of ascidian powder in aquatic feeds: In the preparation of complete aquafeeds, ascidian powdered meal can be substituted for routine fish at 64%, thereby the physical and chemical qualities of the aquafeeds will not be affected. Further, the oil absorption ability and particle expansion rate of the aquafeeds have been reported to be considerably enhanced by the optimal viscosity of the ascidian powdered meal (Gao et al., 2023).

CHAPTER 6

NATURAL COMPOUNDS OF ASCIDIANS AND THEIR THERAPEUTIC VALUES

Among the marine invertebrates, sponges, bryozoans, and ascidians are known to produce a rich diversity of bioactive compounds which largely belong to the chemical classes such as alkaloids, peptides, polyketides, quinones, and steroids. During 1994–2014, about 600 new metabolites have been derived from these ascidians. A total of about 1200 structurally distinct, bioactive compounds have so far been identified from these ascidians with diverse bioactivities including antitumor, antiviral, antimalarial, anti-inflammatory, antibacterial, antiparasitic, and antidiabetic (Watters, 2018; Santhanam and Ramesh, 2019). These secondary metabolites help the animals mainly to avoid predation. In the case of ascidians, many of such metabolites are produced by their endosymbiotic bacteria.

6.1 ASCIDIAN-DERIVED NATURAL COMPOUNDS: THEIR CHEMICAL CLASSES AND BIOACTIVITIES

The percentages of chemical classes of ascidian-derived secondary metabolites and their bioactivities are given in Tables 6.1 and 6.2.

6.2 APPROVED DRUGS FROM ASCIDIANS

Two ascidian metabolites have been marketed for cancer treatment. They are Ecteinascidin (ET-743, trabectedin), an FDA-approved drug (under the trade name Yondelis®) which is derived from the endosymbiotic

Biotechnological Potential of Ascidians: Wacky and Weird Sea Squirts. Ramasamy Santhanam (Author)

bacterium, *Candidatus Endoecteinascidia frumentensis* hosted by the mangrove ascian, *E. turbinata*. The other one is Aplidin® (dehyrodidemnin B, plitidepsin) which was isolated from *Aplidium albicans.* This drug which is currently in phase III clinical trials has been granted orphan drug status by FDA; and it has also been approved by the Australian Regulatory Agency (https://pharmacy.ucsf.edu/news/2021/02/cancer-drug-stops-sars-cov-2-lab; Watters, 2018; Onder and Ay, 2022).

TABLE 6.1 Percentage of Chemical Classes of Ascidian-derived Secondary Metabolites

Chemical Class	**%**
Alkaloids	74
Polyketides	7
Peptides	5
Quinones	3
Steroids	3
Others	8

Source: Adapted from Dou and Dong (2019).

TABLE 6.2 Percentage of Bioactivities of Secondary Metabolites of Ascidians

Bioactivity	**%**
Antitumor	67
Antibacterial	10
Anti-inflammatory	4
Antiviral	3
Anti-HIV	3
Antioxidant	1
Others	12

Source: Adapted from Dou and Dong (2019).

Ecteinascidia turbinata

Image credit: Public domain.

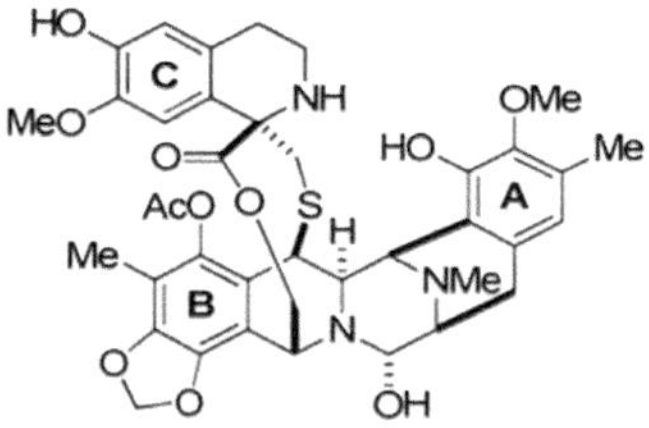

Ecteinascidin 743

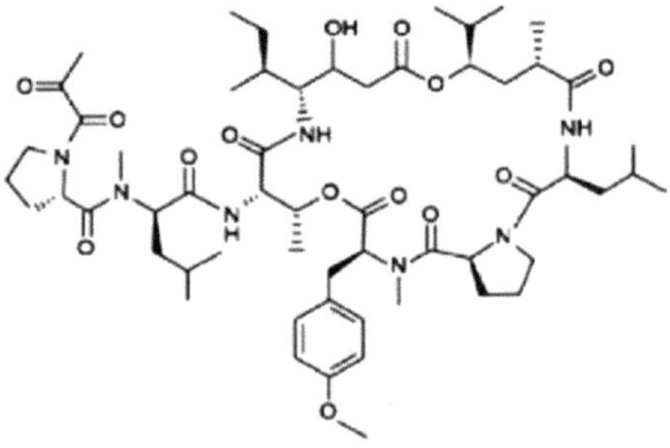

Aplidium sp. Plitidepsin (Dehydrodidemnin B)

Image credit: jkirkhart35. https://creativecommons.org/licenses/by/2.0/

6.3 THERAPEUTIC COMPOUNDS OF ASCIDIANS IN PIPELINE

The therapeutic compounds of ascidians approved as drugs by the FDA and in clinical trials are given in Table 6.3.

TABLE 6.3 Therapeutic Compounds of Ascidians in Pipeline

Compound	NP/D	Source Organism	Chemical Class	Against	Status
Trabectedin (Ecteinascidin 743)	NP	*Ecteinascidia turbinata*	Tetrahydro-isoquinoline alkaloid	Cancer	FDA-approved on Oct. 23, 2015
Plitidepsin (dehydrodidemnin B, Aplidine)	NP	*Aplidium albicans*	Cyclic depsipeptide	Cancer	Phase III/ Orphan drug[a]
Didemnin B	NP	*Trididemnum solidum*	Cyclic depsipeptide	Cancer	Phase III (stopped)
Trabectedin analog (PM01183)	D	Tunicate	NRPS-alkaloid	Cancer	Phase I
Midostaurin	Synthetic analog	Ascidian	Indolocarbazole	Cancer	Phase III(a)
Lestaurtinib (CEP-701)	Synthetic analog	Ascidian	Indolocarbazole	Cancer	Phase III(b)
Enzastaurin (LY317615)	Synthetic analog	Ascidian	Indolocarbazole	Cancer	Phase III(c)
Edotecarin (J-107088)	Synthetic analog	Ascidian	Indolocarbazole	Cancer	Phase III
Becatecarin (XL 119)	Synthetic analog	Ascidian	Indolocarbazole	Cancer	Phase II/III

TABLE 6.3 *(Continued)*

Compound	NP/D	Source Organism	Chemical Class	Against	Status
UCN-01	Synthetic analog	Ascidian	Indolocarbazole	Cancer	Phase II
CEP-1347 (KT7515	Synthetic analog	Ascidian	Indolocarbazole	Parkinson	Phase II
CEP-2563 (prodrug of CEP-751)	Synthetic analog	Ascidian	Indolocarbazole	Cancer	Phase I
Sautosporine (AM-2282)	Synthetic analog	Ascidian	Indolocarbazole	Cancer	Preclinical

Notes: NP/D: Natural product/derivative.

a. *Received orphan drug status for treatment of mastocytosis acute myeloid leukemia (FDA, 2009, 2010), NCI, National Cancer Institute; CTN, clinical trial number.*

b. *Received orphan drug status for AML (FDA, 2007).*

c. *Orphan drug status for diffuse large B-cell lymphoma.*

[a]*Plitidepsin has approved orphan drug status by the European Medicines Agency for treating acute lymphoblastic leukemia.*

Source: Adapted from Palanisamy et al. (2017).

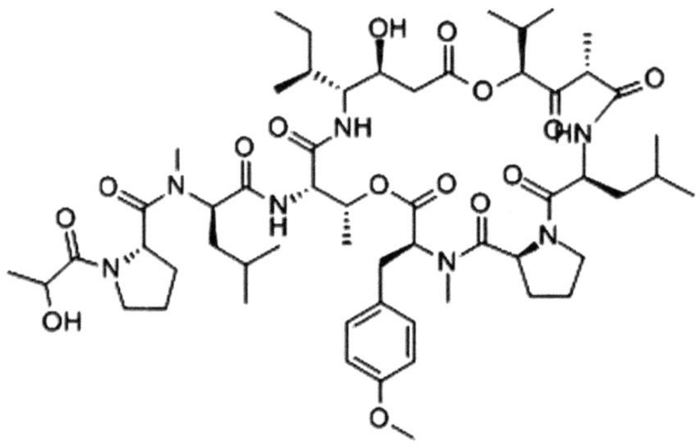

Didemnin B

6.4 THERAPEUTIC COMPOUNDS OF ASCIDIANS AND THEIR BIOACTIVITIES

The species of ascidians possessing antitumor, antiviral, anti-HIV, antimalarial, antidiabetic and antituberculosis activities are listed in Tables 6.4 and 6.5.

TABLE 6.4 Antitumor, Anticancer and Antiproliferative Activities of Ascidians (1)

Species	Compound(s)	Chemistry	Against	IC50 (μM)
Aplidium albicans	*Plitidepsin* (generic name aplidine or dehydrodidemnin B; tradename aplidin)	Cyclic depsipeptide	Multiple myeloma, lymphoblastic leukemia	–
Aplidium altarium	Botryllamide K, L	Tyrosine derivatives	MCF-7, H460	74.9; 91.9
Aplidium caelestis	Caelestine A–D	Brominated quinolinecarboxylic acids	MCF-7, MM96L	39, 49, 40, 38 μM; 62, 69, 54, 52
Aplidium confluata	Aplidiopsamine A	Alkaloid	HEK-293	120
Aplidium conicum	Conicaquinone A, B	Alkane derivatives	Human leukemia	3
-do-	Conicaquinone A	-do-	MCF-7	44.5
Aplidium cyaneum	Aplicyanin B, D	Indole alkaloids	A-549, HT-29, MDA-MB 231	0.66, 0.63; 0.39, 0.33; 0.42, 0.41
Aplidium glabrum	Haouamine A	Alkaloid	HT-29	0.1
-do-	Haouamine B	-do-	MS-1	5.0
-do-	Extract	–	COAD	5.8
Aplidium longithorax	Longithorone A	Prenylated quinone	P-388	ED50 > 10 μg/mL
-do-	Floresolide A–C	Synthetic	KB	1–10
Aplidium meridianum and *Aplidium falklandicum*	Analogs of meridianins A-G	Brominated 3-(2-aminopyrimidine)-indoles	Antitumor	(2)
Ascidia mentula	3,7,11,15-Tetramethyl-hexadecan-1,19-sodium disulfate; heneicosane-1,21-sodium disulfate	Alkyl sulfates	IGR-1	> 110, 100
Botryllus tyreus	Botryllamide E, F, G	Bromotyrosine derivatives	Breast cancer carcinoma	27, 41, 7

TABLE 6.4 *(Continued)*

Species	Compound(s)	Chemistry	Against	IC50 (µM)
Ciona savignyi	Chondromodulin-1	Glycoprotein	Human neuroblastoma and cervical cancer	– (3)
-do-	CS5931	-do-	Several tumor cell lines	– (4)
Clavelina phlegraea	Clavaminol A	2-Amino-3-alkanol	AGC	5
Cnemidocarpa bicornuta	2-(3′-Bromo-4′-hydroxyphenyl) ethylamine	Brominated alkaloid	P-388	46
Cystodytes dellechiajei	Sebastianine A, B	Alkaloids	HCT-116	5.1; 0.9
Cystodytes violatinctus	Arnoamine D	Pyridoacridine alkaloid	HCT-116, SW480, A375	4.3; 8.5; 6.0
Cystodytes sp.	Arnoamine A	-do-	MCF-7, A-549, HT-29	0.3; 2.0; 4.0
-do-	Arnoamine B	-do-	-do-	5; 2; 3
Diazona angulata	Diazonamide A	Peptide	CA46, MCF7, PC3, A549	3.8, 1.9, 2.3, and 4.5 nm (5)
Didemnum cuculiferum	Vitilevuamide	13 Amino acid peptide	Human tumor cell lines	6–311 nM (6)
Didemnum granulatum	Granulatimide, isogranulatimide	Alkaloids	p53	1–8
Didemnum lahillei	Crude compound	–	Colon adenocarcinoma; small cell lung carcinoma	33; 49
Didemnum molle	Mollamide B	Alkaloid	H460, SF-268, MCF7	29; 42; 44
-do-	Mollamide C	-do-	L1210, MCF7	–
-do-	Mayotamide, A, B; Comoramide A, B	Cyclic heptapeptides	A549, HT29, MEL-28	5–10

TABLE 6.4 *(Continued)*

Species	Compound(s)	Chemistry	Against	IC50 (µM)
Didemnum obscurum	Lamellarin-ζ, lamellarin-χ, lamellarin-F and lamellarin-T diacetate	Alkaloids	COLO-205	–
Didemnum proliferum	Shishijimicin A-C	Enediyne antibiotics	HeLa	1.8–6.9 pM (12)
Didemnum psammatodes	Methyl myristate, methyl palmitate and methyl stearate	Methyl esters	HL-60, K-562, CEM, Molt-4	9.5, 6.9, 14.3; 9, 5.6, 14.3; 10, 4.3, 22.9; 2.4,1.6,3.6
Didemnum rubeum	Plakinidine D	Pyrroloacridine alkaloid	HCT-116	5
Didemnum sp. 1	Ascididemin	Pyrido [2, 3, 4-kl] acridine-based alkaloid	P-388, HCT116, MCF-7	0.4; 0.3;0.3
Didemnum sp. 2	Didemnoline A–C	β-Carboline	KB	6.1; 3; 0.3
Didemnum sp. 3	3-Bromofascaplysin	Alkaloid	HCT-116	–
Diplosoma virens	Virenamide A–C	Tripeptides	P388; A549, HT29, CV1	2.5; 10
Distaplia skoogi	3,6-Dibromoindole, 6-bromo-3-chloroindole, 6-bromo-2-oxindole	Indoles	MDA-MB-231	117.7, 72.5, 74.4
Ecteinascidia turbinata (its endosymbiotic bacterium, '*Candidatus Endoecteinascidia frumentensis*'	Ecteinascidin 743	Tetrahydroisoquinoline	L1210, P388, A549, HT29, MEL28, CV-1	5.0; 0.2; 0.2; 0.5; 5.0; 1.0 nM (11)
Eudistoma gilboverde	14-Methyleudistomidin C	β-Carboline alkaloid	LOX, COLO-205, MOLT-4, OVCAR-3	0.41, 0.42, 0.57, 0.98

TABLE 6.4 *(Continued)*

Species	Compound(s)	Chemistry	Against	IC50 (μM)
Eudistoma glaucus	Eudistomidin J	-do-	P388, L1210, KB	– (14)
Eudistoma toealensis	3-*OH*-staurosporine	Indolocarbazole alkaloid	MONO-MAC-6	13 nM
Eudistoma vannamei	2-Hydroxy-7-oxostaurosporine, 3-hydroxy-7-oxostaurosporine	Staurosporine derivatives	Jurkat, Molt-4	10.3 nM, 18.6 nM
Eudistoma sp.	7β-Hydroperoxycholesterol	Hydroperoxysterol	Hep3B, A549	15.6; 17.4
-do-	7α-Hydroperoxycholesterol	-do-	MCF7, MDA-MB-231	11.2; 11.2
Halocynthia papillosa	(*R*)-2,6-Dimethylheptyl sulfate, 6-methylheptyl sulfate, (*E*)-5-octenyl sulfate	Sulfated alkene and alkanes	WEHI	20.9, 15.0, 12.2
Halocynthia roretzi	Aqueous extracts of edible parts	–	HepG-2	– (15)
Halocynthia roretzi	Hydrolysates	–	HT-29	– (16)
Herdmania momus	MeOH extracts	Amino acid derivatives	Ac2F rat liver cells	PPAR-γ activation (7)
Leptoclinides dubius	*N*-(1*H*-indolyl-3-carbonyl)-D-arginine, *N*-(6-bromo-1*H*-indolyl-3-carbonyl)-l-arginine, *N*-(6-bromo-1*H*-indolyl-3-carbonyl)-l-histidine	Indoles	P-388, KB	–
-do-	Leptoclinidamide, C 2-α-D-mannosylpyranosyl-L-tryptophan	Tryptamine-derived alkaloids	HCT-15, Jurkat	–
Lissoclinum badium	Lissoclibadin 1–3	Polysulfur alkaloids	HL-60	0.4, 0.2, 5.5

TABLE 6.4 *(Continued)*

Species	Compound(s)	Chemistry	Against	IC50 (µM)
-do-	Lissoclibadin 2, lissoclibadin 3 (=lissoclinotoxin E)	-do-	VL 79	0.8; 0.3
-do-	Lissoclibadin 2	-do-	DLD-1, HCT116	2.9, 2.5
-do-	-do-	-do-	MDA-MB-231, ACHN, NCI-H460	0.2,; 0.2; 0.4
Lissoclinum notti	Diplamine, lissoclinidine	Pyridoacridine alkaloids	P388 D1	1.9, 4.6
-do-	-do-	-do-	HCT-116	<1.4; 3
-do-	*N*-(2-(6-oxo-6*H*-pyrido[2,3,4-kl] acridin-4-ylamino)ethyl)pyrazine-2-carboxamide; 2-(6-oxo-6*H*-pyrido[2,3,4-kl]acridin-4-ylamino) ethyl pyrazine-2-carboxylate	Alkaloids	P388	>25
Lissoclinum sp.	Haterumaimide O, P	Labdane alkaloids	P388	0.23 nM; 0.45 nM
-do-	Dichlorolissoclimide, chlorolissoclimide	-do-	P388	4.1 nM; 5.5 nM
-do-	Mandelalide A, B	Polyketide macrolides	NCI-H460	12 nM; 44 nM
Microcosmus vulgaris	Sulcatin	*N*-methylpyridinium alkaloid	Murine tumor cell lines	3.65 µM (17)
Molgula manhattensis	Crude extract	–	HT-29	5.8
Perophora namei	Perophoramidine	Alkaloid	HCT-116	60
Phallusia fumigata	Phalluside 1,2	Glucosphingolipids	HT29	10
Polycarpa clavata	Polycarpine dihydrochloride	Alkaloid	HCT-116	0.9
Polysyncranton lithostrotum	Vitilevuamide	13 amino acid peptide	Human tumor cell lines	6–311 nM (6)

TABLE 6.4 *(Continued)*

Species	Compound(s)	Chemistry	Against	IC50 (µM)
-do-	Namenamicin	Enediyne antibiotic	Cytotoxicity	3.5 nM (13)
Pseudodistoma cereum	1,3,7-Trimethylisoguanine	Purine	P-388, SF-268	–
Stolonica sp.	Stolonic acid A, B	Unsaturated fatty acids	Human melanoma and ovarian tumor	0.05–0.1 µM
Syncarpa oviformis	Ethyl indolyl-3-glyoxyiate, ethyl 6-bromoindolyl-3 glyoxylate, ethyl 6-bromo-5-hydroxyindolyl-3-glyoxylate	Indole alkaloid derivatives	Ehrlich carcinoma	61; 35; 97
Synoicum adareanum	Palmerolide D-G	Polyketides	Human melanoma	–
-do-	Hyousterones A–B; abeohyousterone	Ecdysteroids	HCT-116	10.7; 3.7; 3.0
Synoicihabitans palmerolidicus	Palmerolide A	-do-	-do-	– (8)
Synoicum pulmonaria	Synoxazolidinone A	Alkaloid	HT-29	30.5
Synoicum sp.	Prunolide A	Rubrolide derivative	Breast cancer	–
Tridideмnum Cyclops	39-oxobistramide K	Lipopeptide	A2780	0.34
Unidentified sp.	Coproverdine	Alkaloid	P388; A549	1.6; 0.3
-do-	-do-	-do-	HT29, MEL28, DU145	0.3
-do-	Lamellarin N	Pyrolle alkaloid	SK-MEL-5; UACC-62	1.9; 9.9

MCF-7, human breast cancer; H460, human nonsmall cell lung cancer; MM96L, human melanoma; HEK-293, human embryonic kidney 293; A-549, lung carcinoma; HT-29, human colorectal adenocarcinoma; MDA-MB 231, human breast cancer; MS-1, human lung carcinoma; COAD, colon adenocarcinoma; KB, epidermoid carcinoma; IGR-1, human melanoma; AGC, Atypical glandular cells; P-388, murine leukemia; HCT-116, human colorectal carcinoma; SW480, human colorectal adenocarcinoma; A375, human melanoma; CA46, B-lymphocyte cell; PC3, human prostate cancer; p53, breast cancer; SF-268, CNS cell line; L1210, murine leukemia; MEL-28, human melanoma; COLO-205, colorectal cancer; HL-60, promyeloblastic leukemia; K-562, myelogenic leukemia; CEM, lymphoblastic leukemia; Jurkat and Molt-4, T-cell leukemia; CV1, Monkey kidney cell line; LOX, melanoma; OVCAR-3, ovarian cancer; MONO-MAC-6, human monocytic cell line; Hep3B, human hepatoma; WEHI, murine fibrosarcoma; HCT-15, colon cancer; VL 79, Chinese hamster lung fibroblast; DLD-1, colon cancer; AHN, renal cancer; NCI-H460, large-cell lung cancer; P388 D1, murine leukemia; NCI-H460, human lung cancer; A2780, ovarian cancer cell line; HeLA, human epithelial adenocarcinoma cell line; HepG-2, human liver cancer cell line.

(1) Palanisamy et al. (2017); (2) Núñez-Pons et al. (2015); (3) Dou et al. (2018); (4) Zhao et al. (2013); (5) Cruz-Monserrate et al. (2003); (6) Edler et al. (2002), (7) Li et al. (2012b); (8) Noguez et al. (2011); (9) https://www.wipo.int/wipo_magazine/en/2009/01/article_0008.html; (10) https://pharmacy.ucsf.edu/news/2021/02/cancer-drug-stops-sars-cov-2-lab; (11) Le et al. (2015); (12) Oku et al. (2003); (13) McDonald et al. (1996); (14) Suzuki et al. (2011); (15) Zhu et al. (2022); (16) Oh et al. (2019) and (17) Aiello et al. (2000).

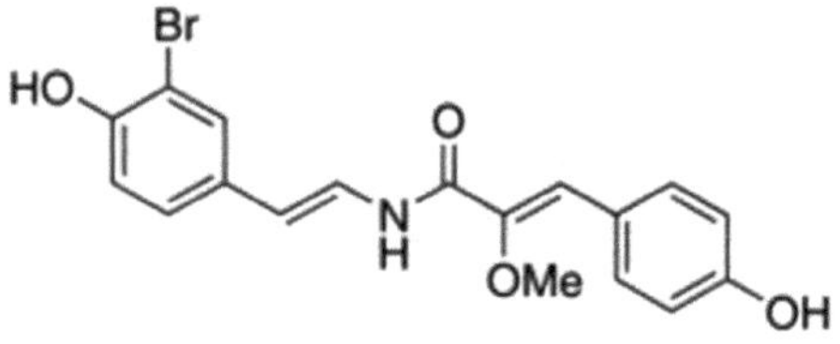

Botryllamide K

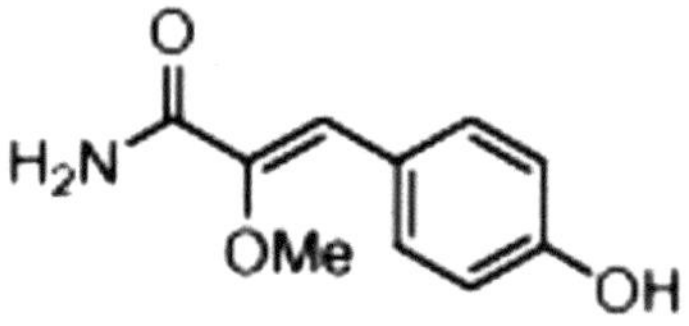

Botryllamide L

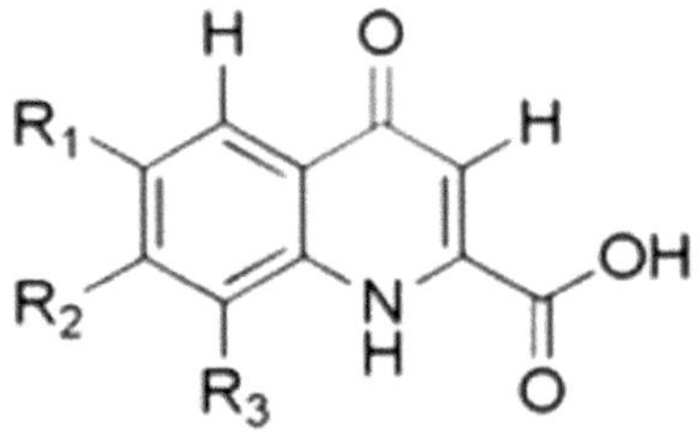

R_1 = H R_2 = Br R_3 = H
R_1 = Br R_2 = Br R_3 = H
R_1 = Br R_2 = H R_3 = OMe
R_1 = Br R_2 = Br R_3 = OMe

Caelestine A–D

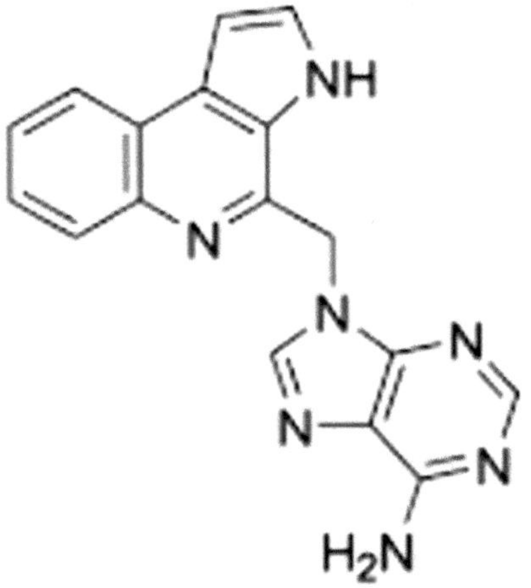

Aplidiopsamine A

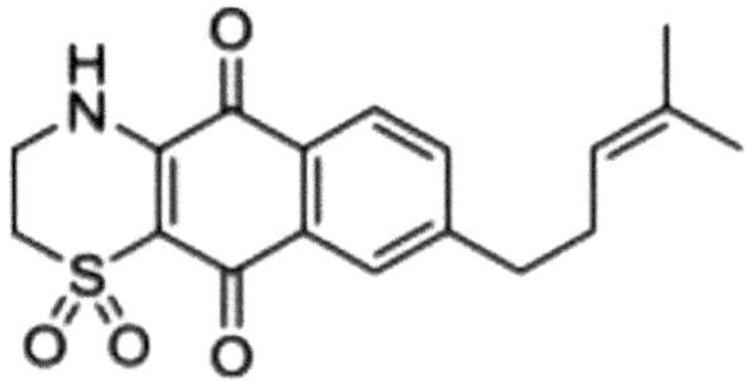

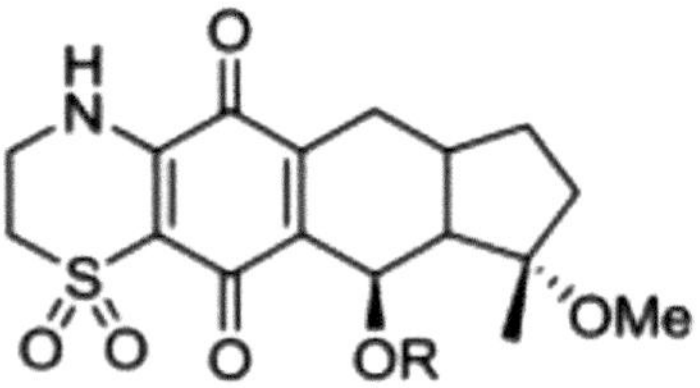

(R=Me)

Conicaquinone A, B

Conicaquinone B

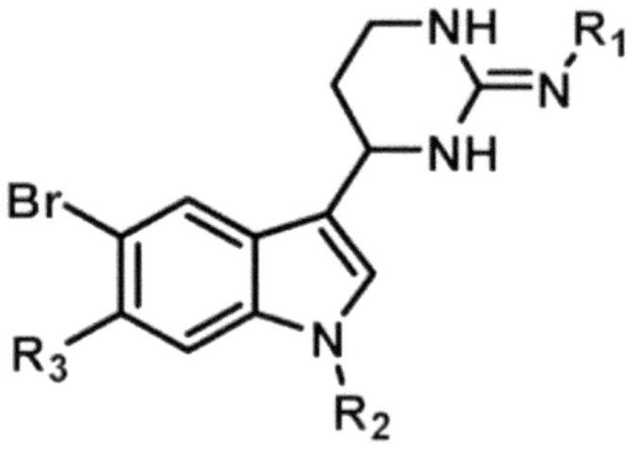

Aplicyanin B (R1 = Ac; R2 = R3 = H).

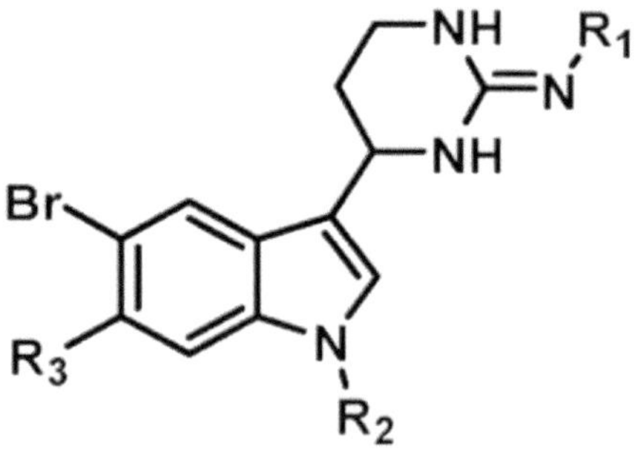

Aplicyanin D (R1 = Ac; R2 = OMe; R3 = H)

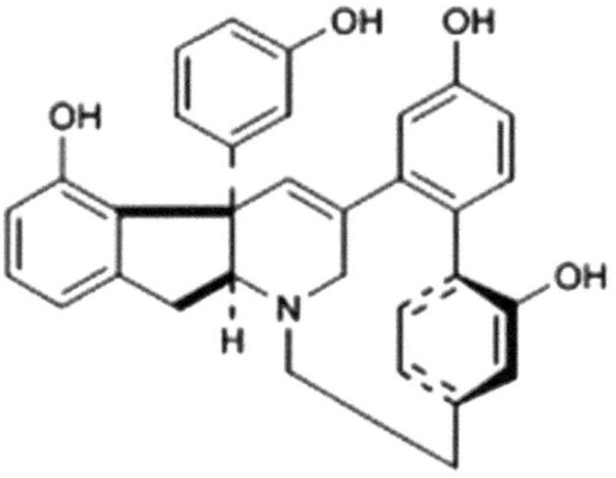

Haouamine A

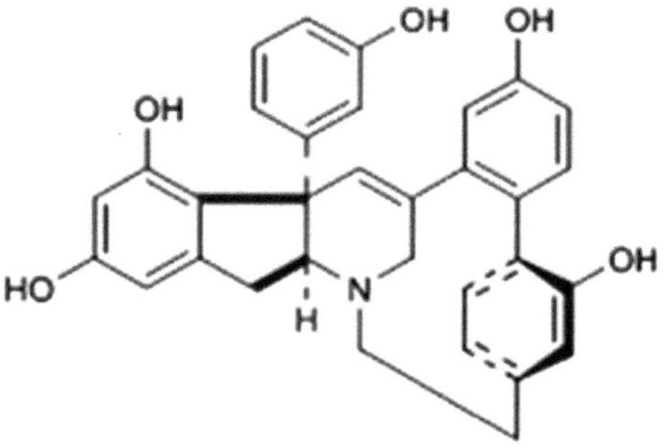

Haouamine B

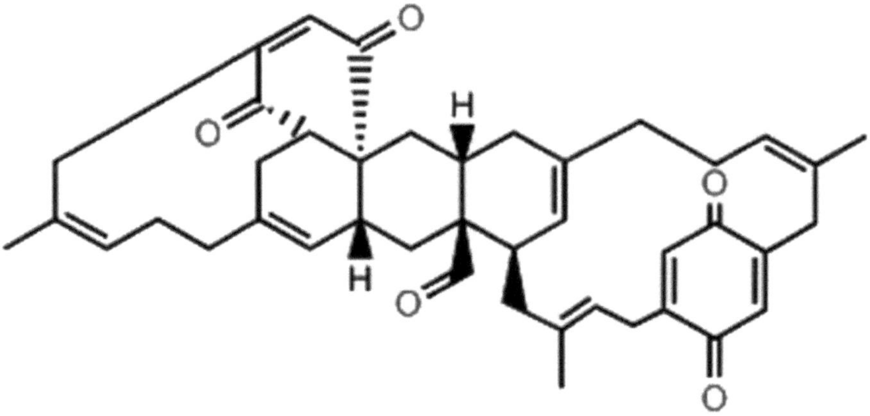

Longithorone A

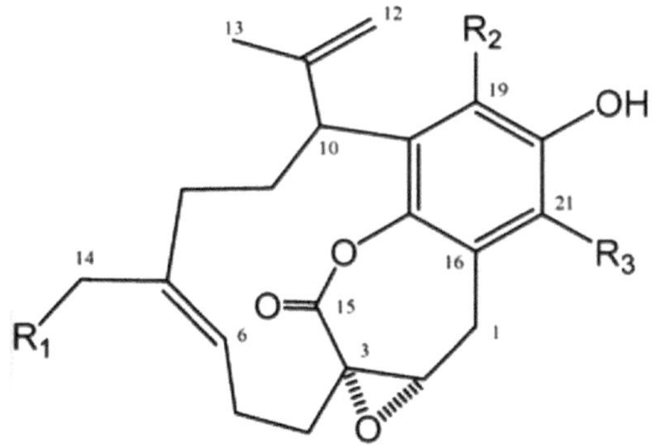

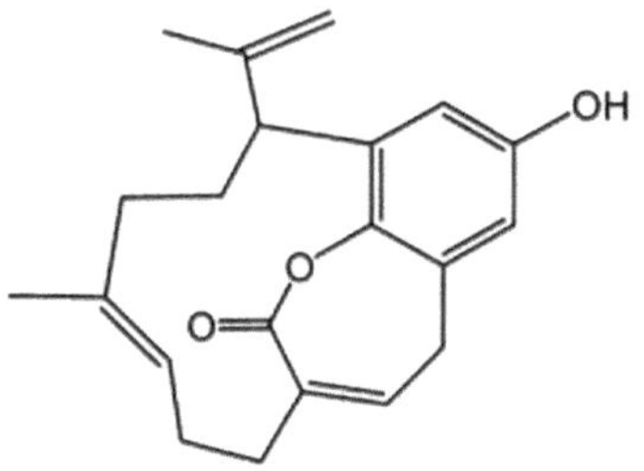

Floresolide A = R1 = R2 = R3 = H

Floresolide B

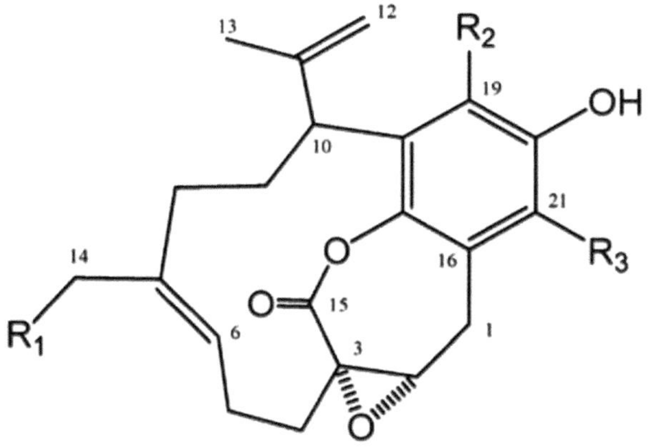

Floresolide C (R1 = OH; R2 = R3 = Br)

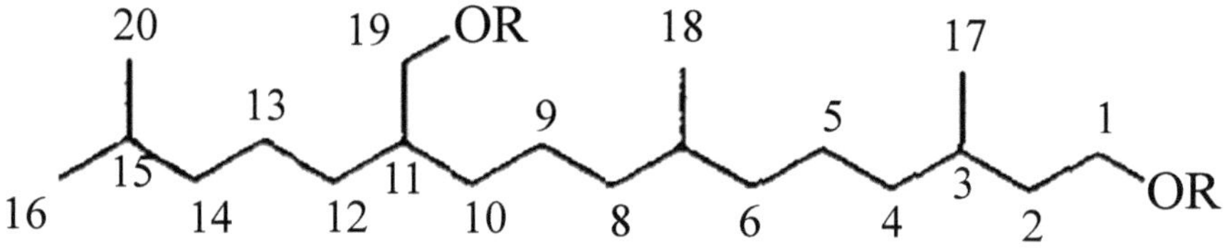

R = $SO_3^-Na^+$

3,7,11,15-Tetramethyl-hexadecan-l,19-sodium disulfate

R = $SO_3^-Na^+$

Heneicosane-l,21-sodium disulfate

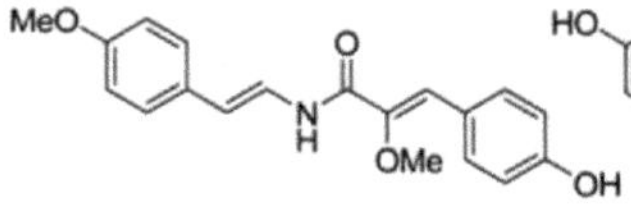

Botryllamide E

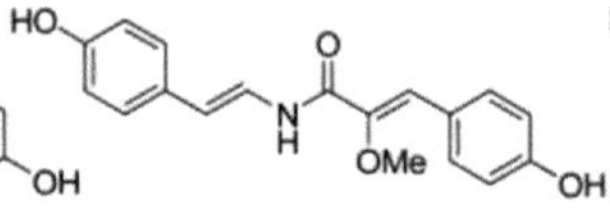

Botryllamide F

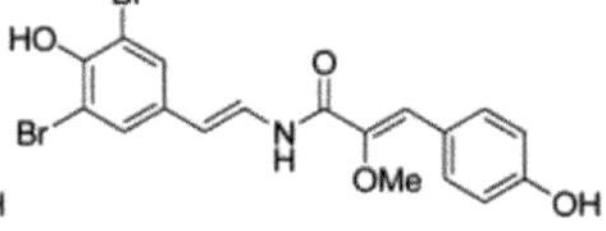

Botryllamide G

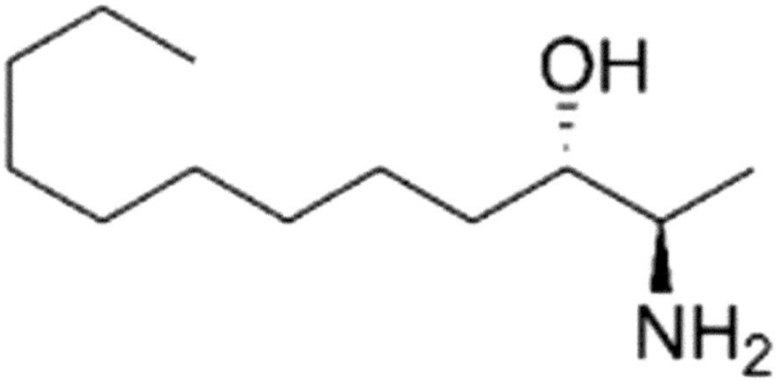

Clavaminol A

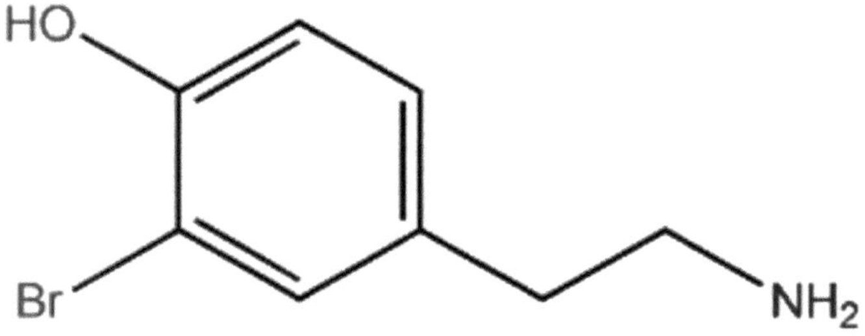

2-(3′-bromo-4′-hydroxyphenyl) ethylamine

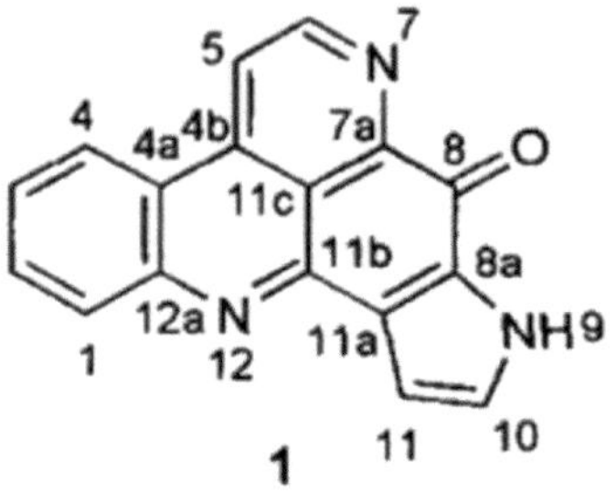

Sebastianine A

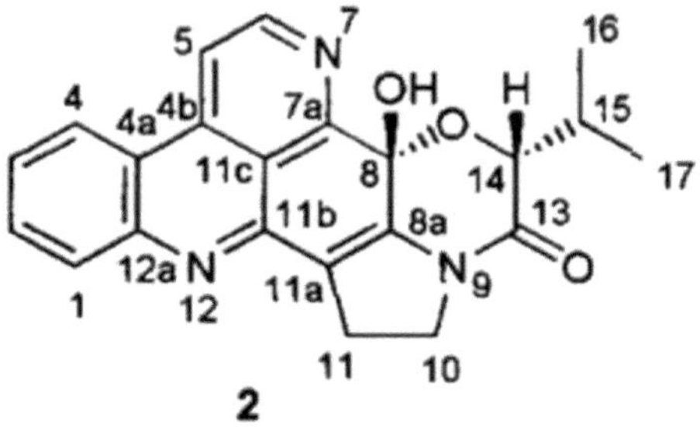

Sebastianine B

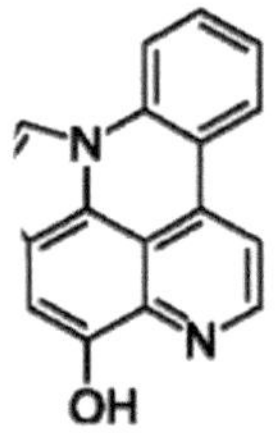

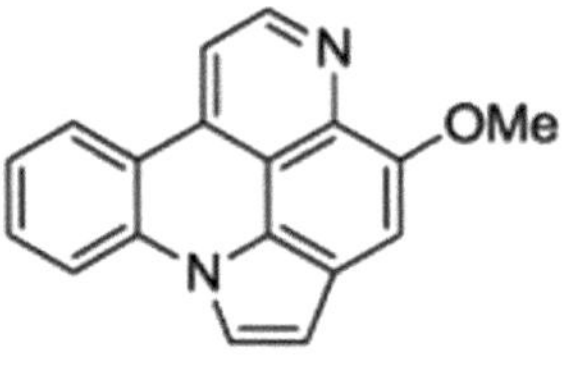

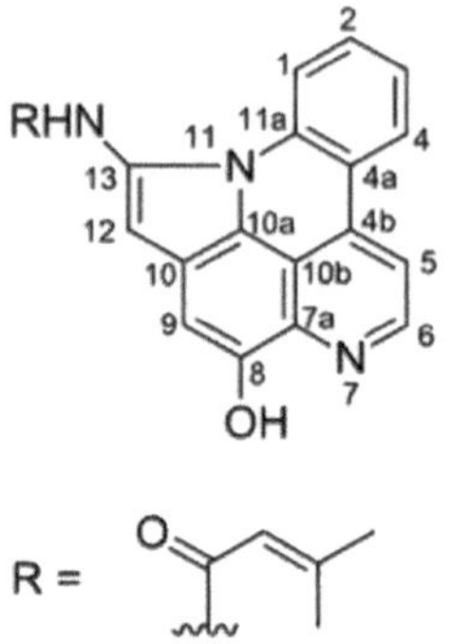

Arnoamine A Arnoamine B Arnoamine D

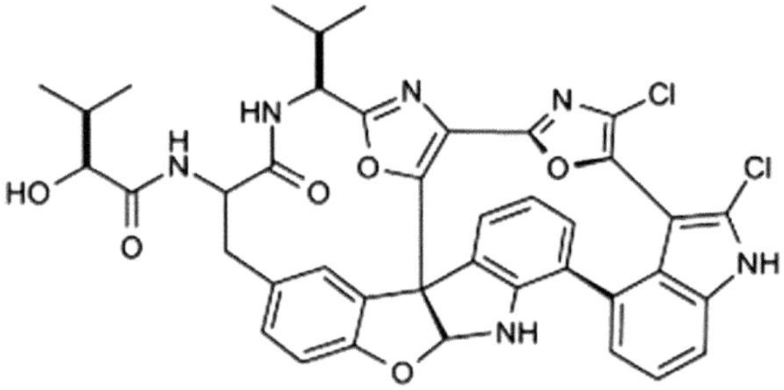

Diazonamide A

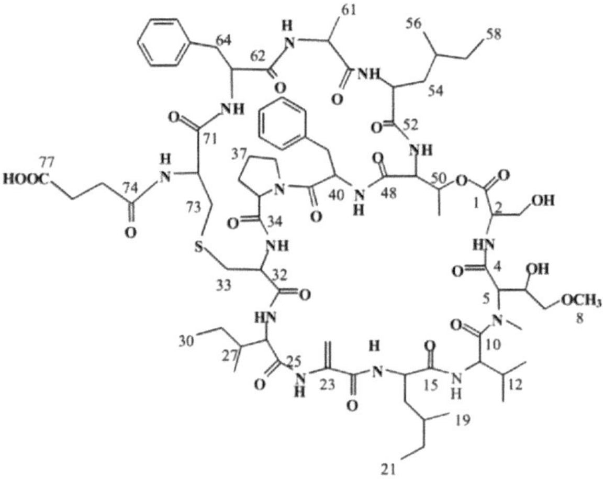

Vitilevuamide

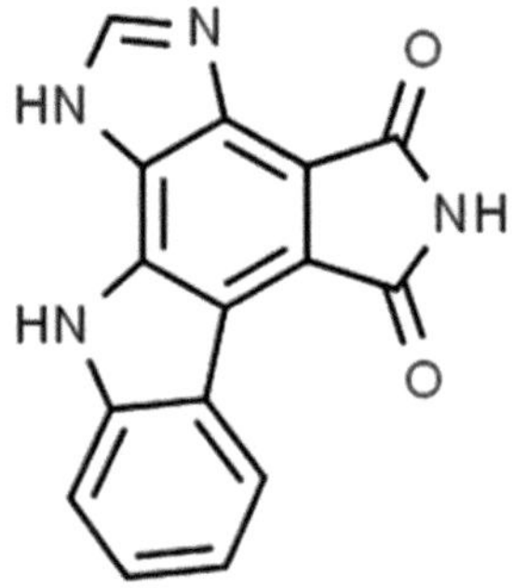

Granulatimide

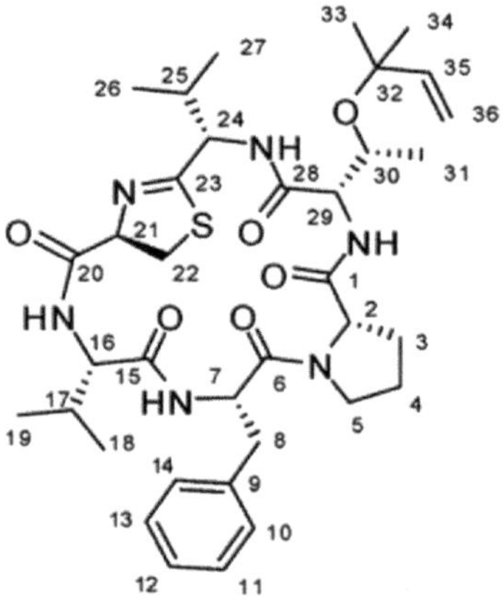

Mollamide B

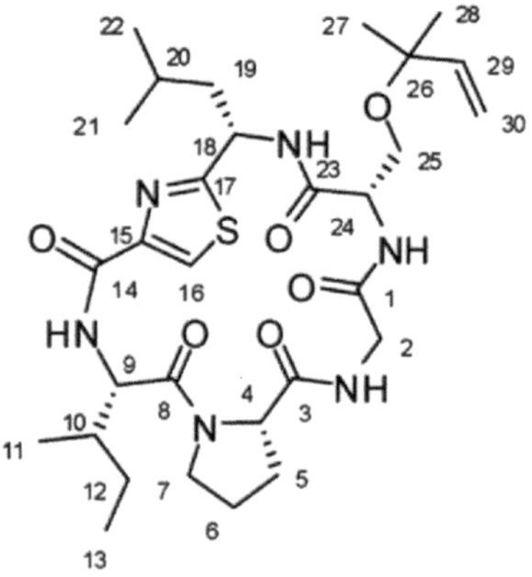

Mollamide C

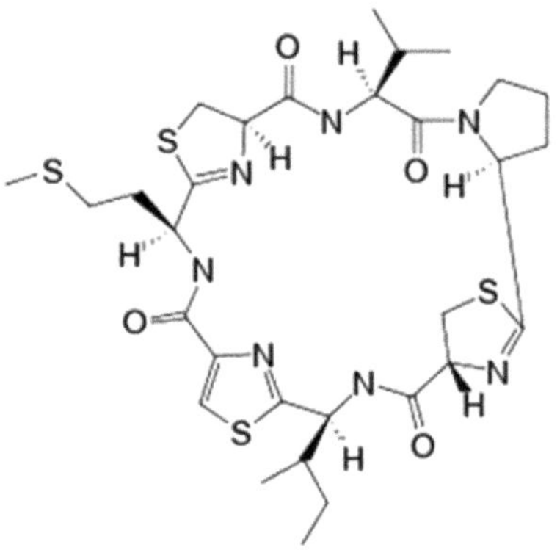

Mayotamide A

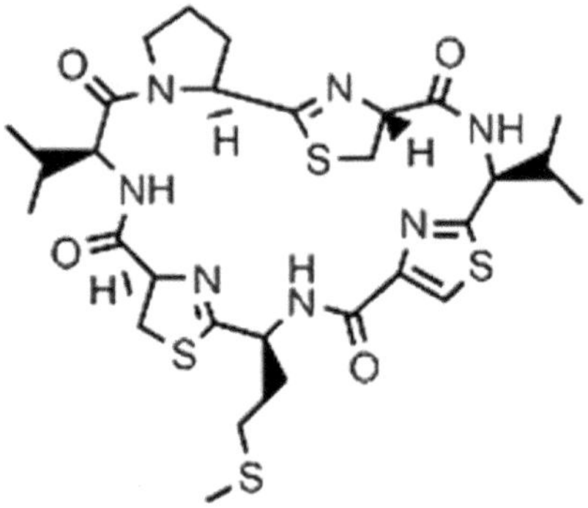

Mayotamide B

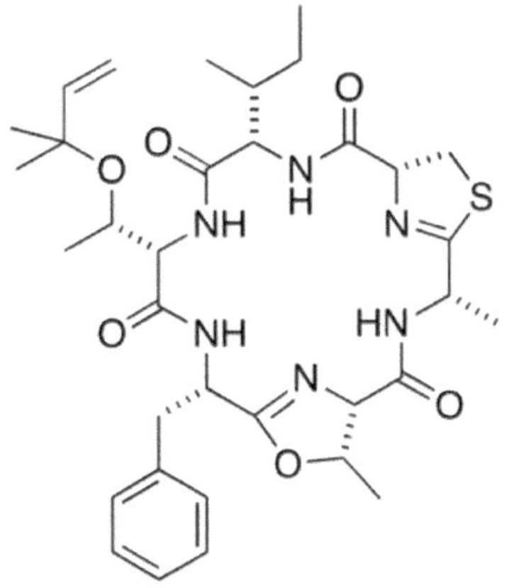

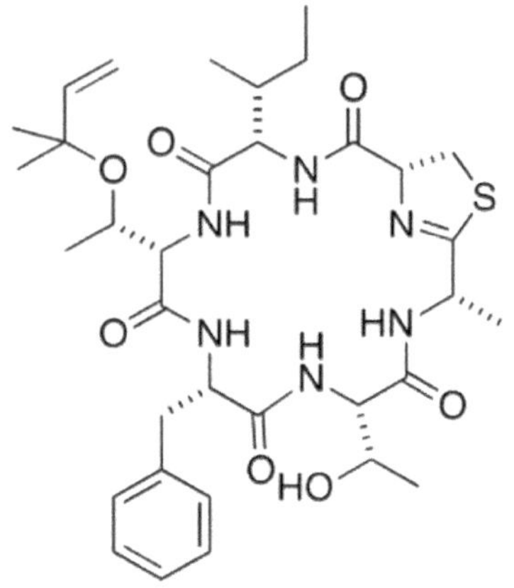

Comoramide A Comoramide B

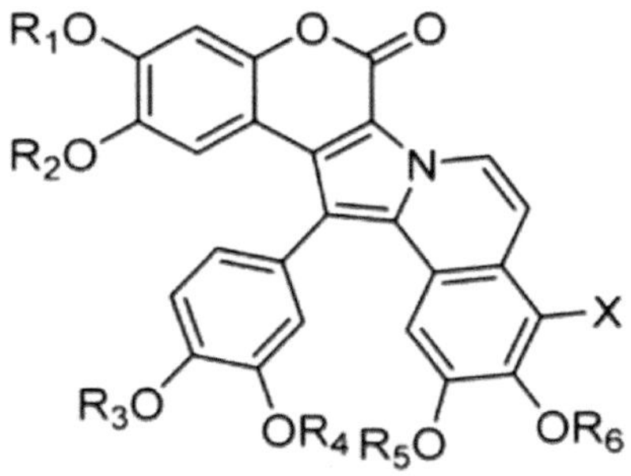

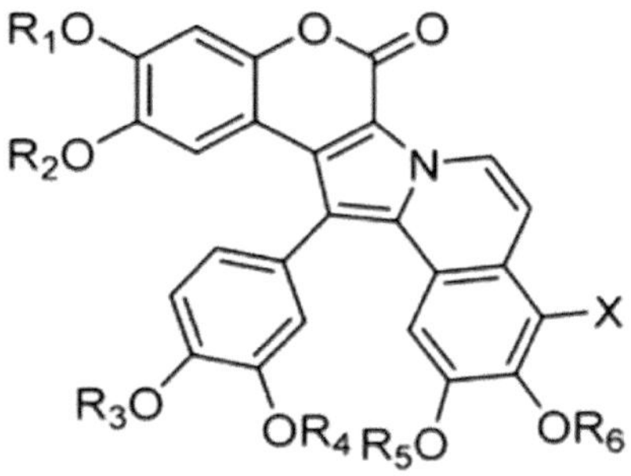

R1 = H; R2= Me; R3 = Me; R4 = Me;
R5 = Me; R6 = Me; X = OMe

R1= H; R2 = Me; R3 = Me; R4= H;
R6 = Me; X = OH

Lamellarin-ζ Lamellarin-χ

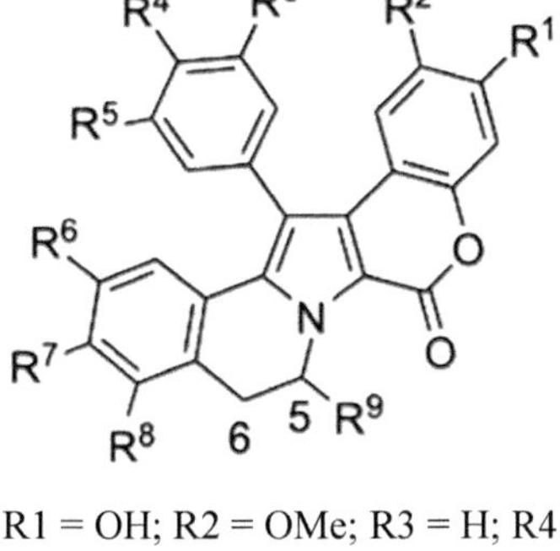

R1 = OH; R2 = OMe; R3 = H; R4 = OMe; R5 = OMe; R6 = OMe; R7 = OMe; R8 = OH;
R9 = H

Lamellarin-F

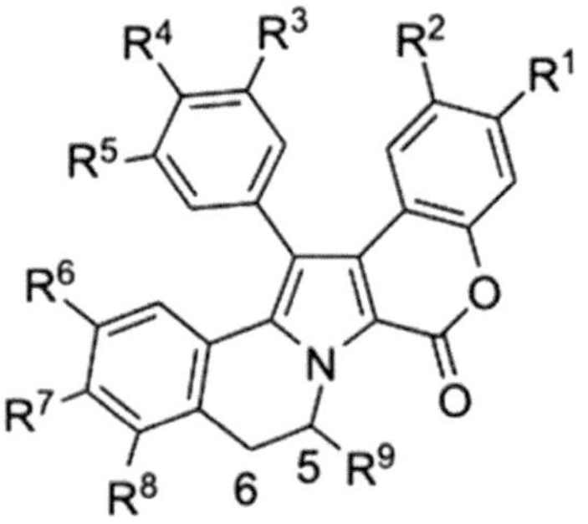

R1 = OAc; R2 = OMe; R3 = H; R4 = OMe; R5 = OAc; R6 = OMe; R7 = OMe; R8 = OMe; R9 = H

Lamellarin-T diacetate

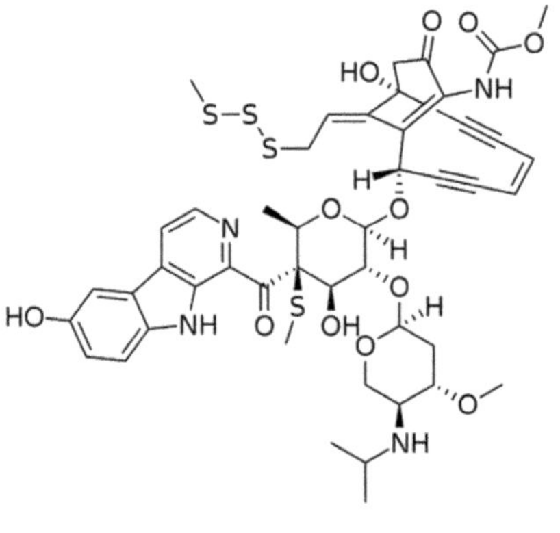

Shishijimicin A

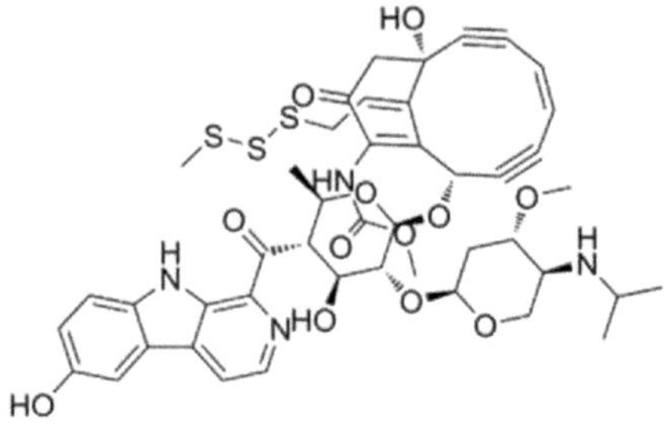

Shishijimicin B

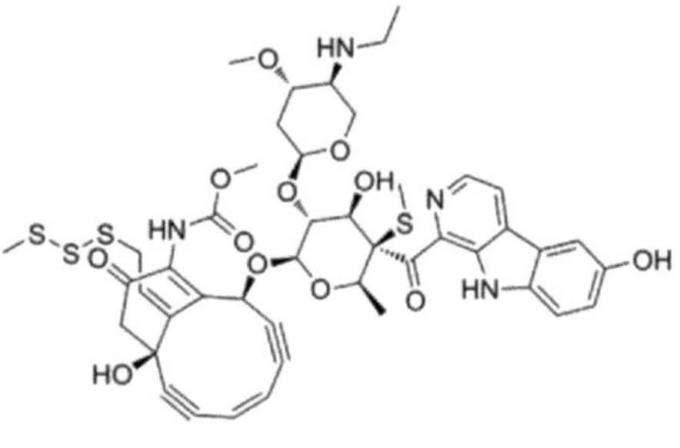

Shishijimicin C

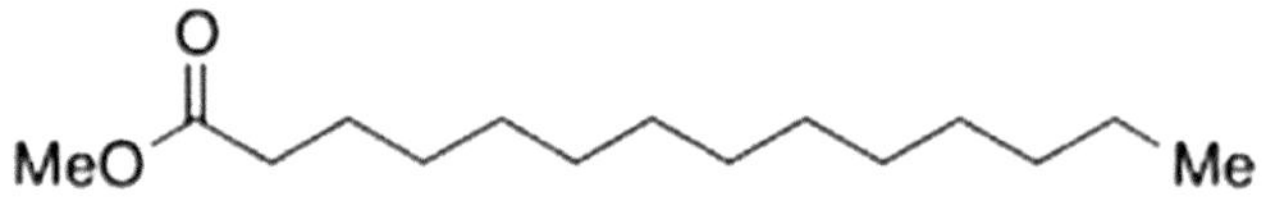

Methyl myristate

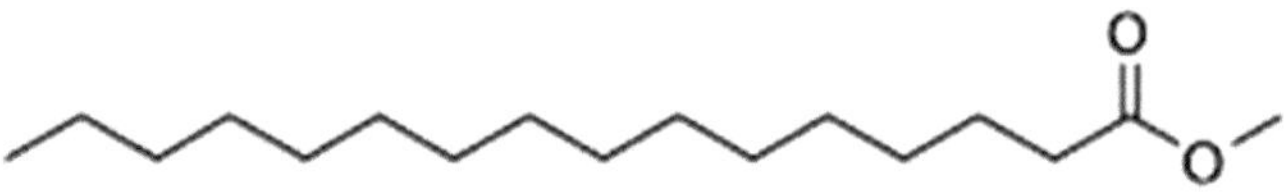

Methyl palmitate

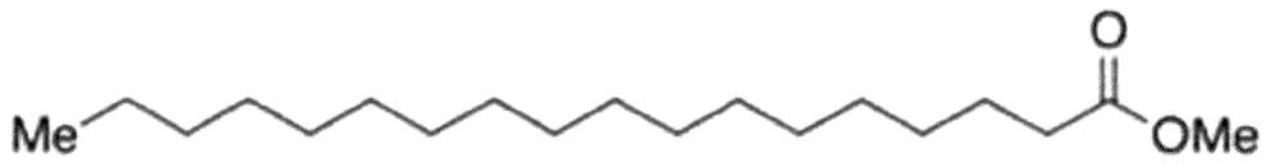

Methyl stearate

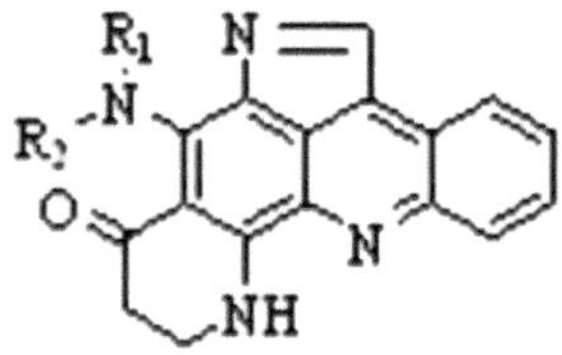

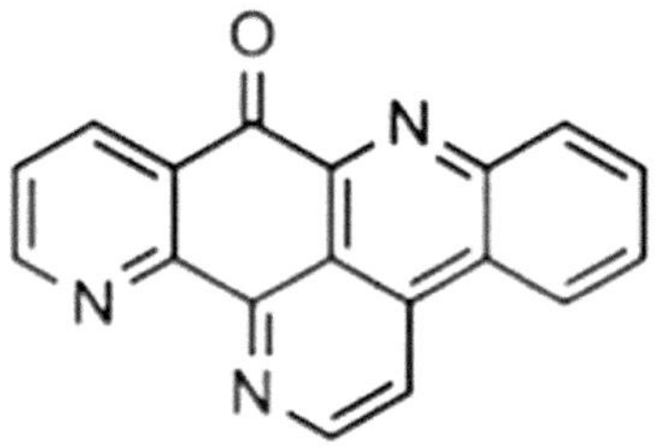

R1 = H; R2 = H

Plakinidine D

Ascididemin

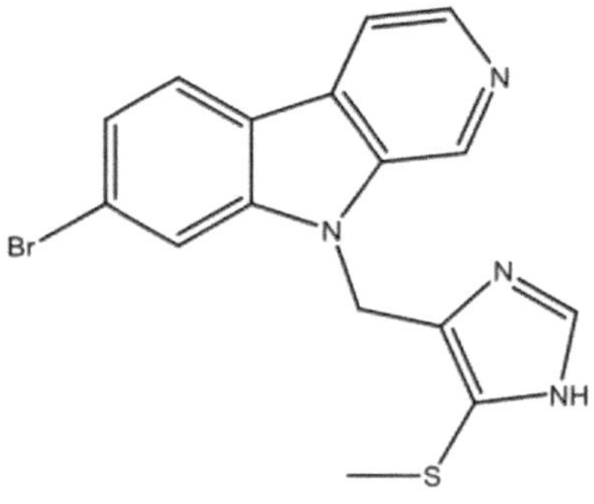

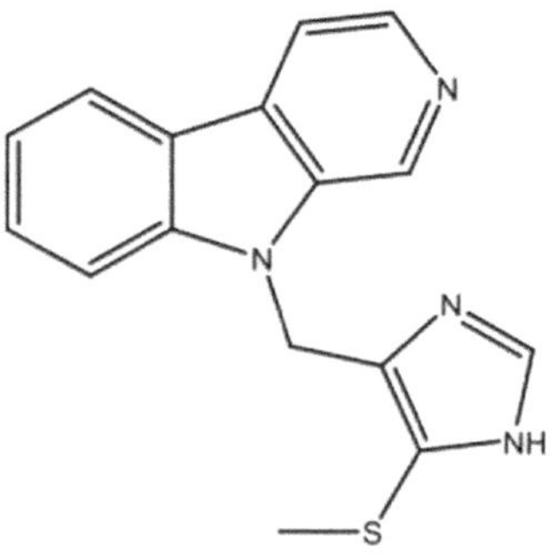

Didemnoline A

Didemnoline B

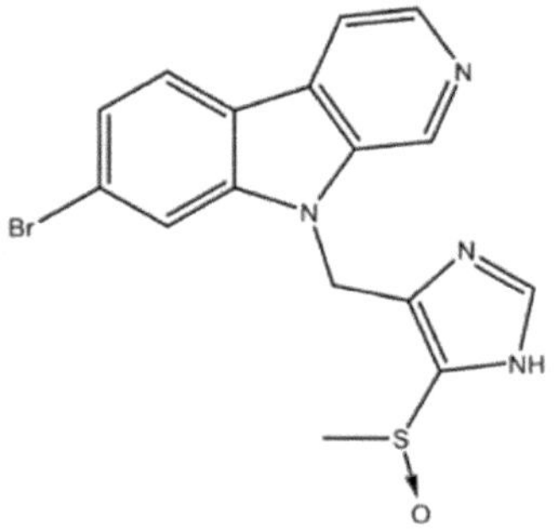

Didemnoline C

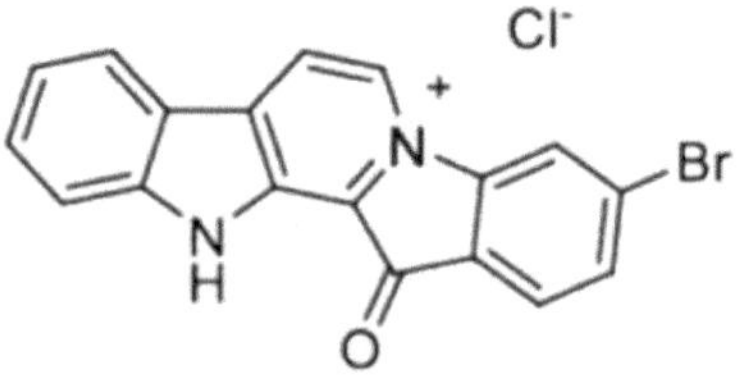

3-Bromofascaplysin

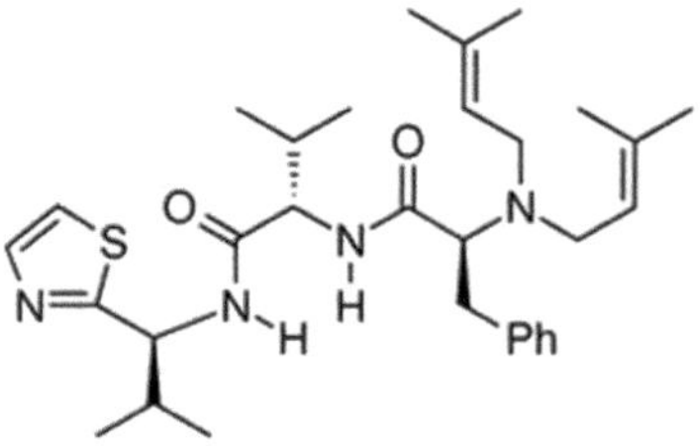

Virenamide A

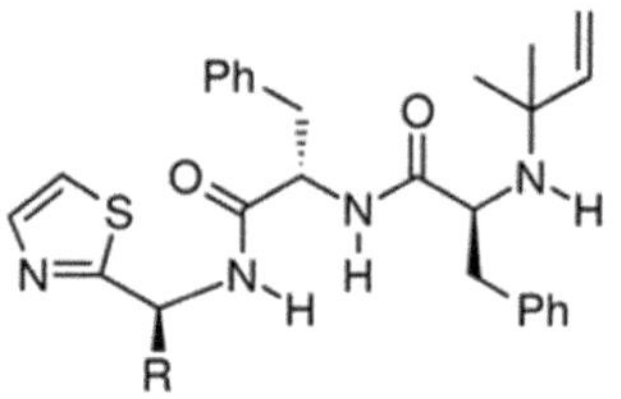

Virenamide B: R = $CH(CH_3)_2$
Virenamide C: R = CH_2Ph

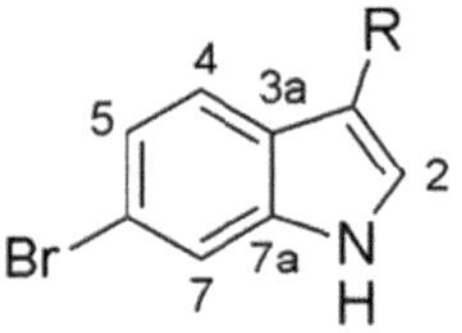

3,6-Dibromoindole
(R–Br)

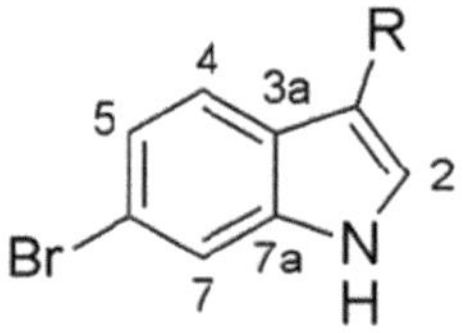

6-Bromo-3-chloroindole
(R = Cl)

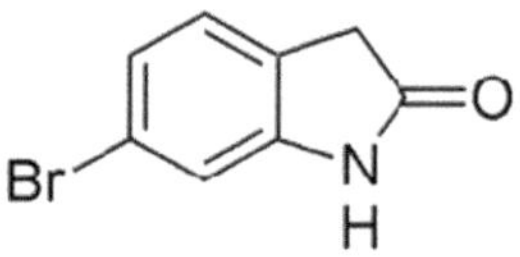

6-Bromo-2-oxindole

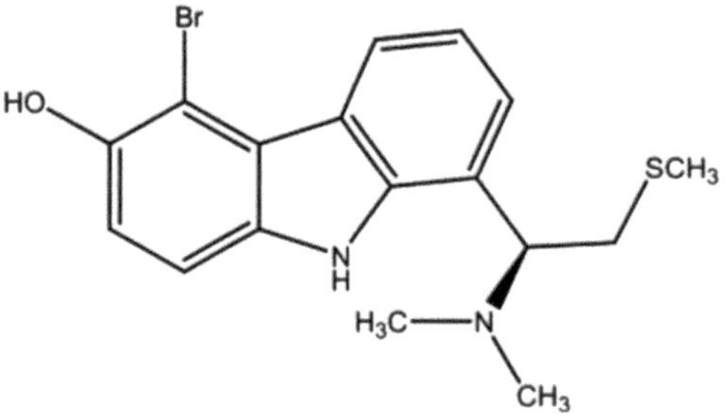

14-Methyleudistomidin C

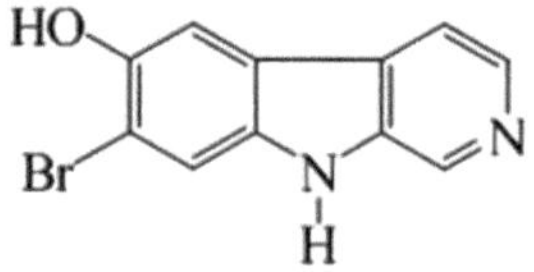

Eudistomidin J

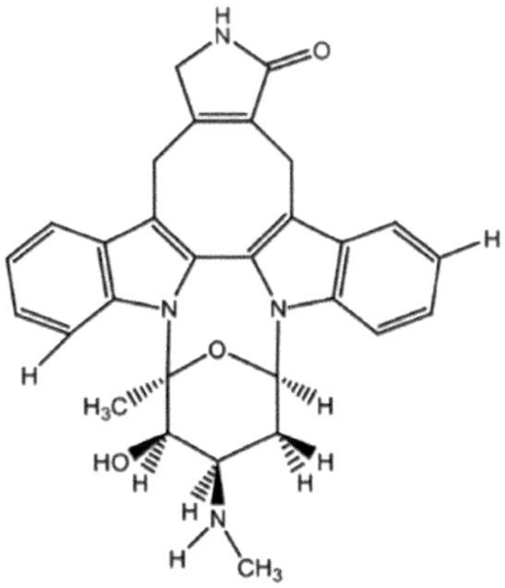

3-*OH*-staurosporine

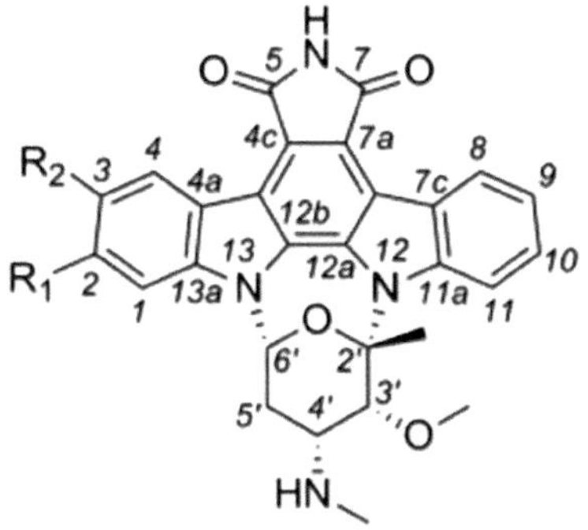

2-hydroxy-7-oxostaurosporine (R1 = OH, R2 = H)
3-hydroxy-7-oxostaurosporine (R1 = H, R2 = OH)

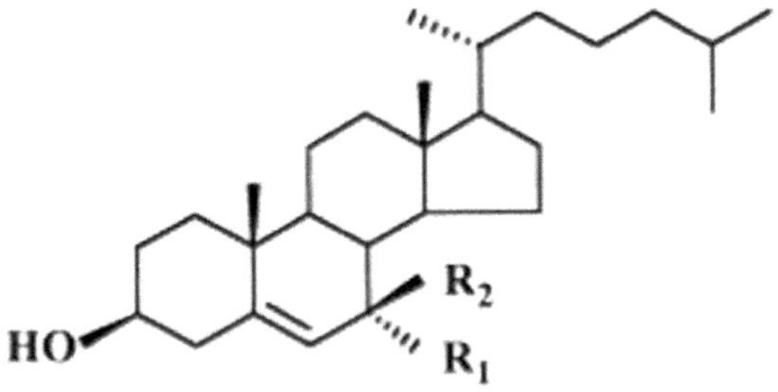

7α-hydroperoxycholesterol (R1 = OOH, R2 = H)

7β-hydroperoxycholesterol (R1 = H, R2 = OOH)

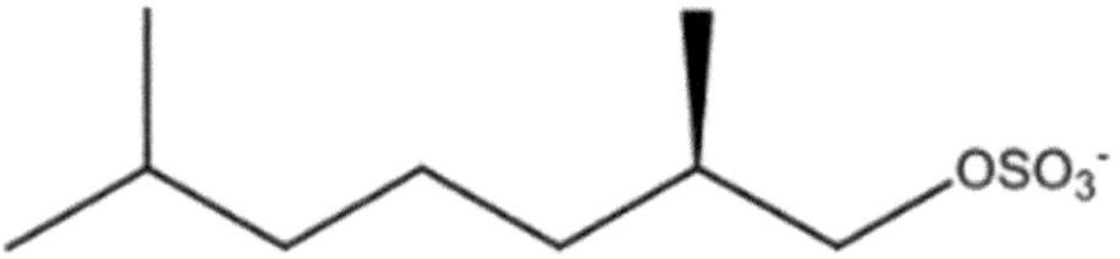

(*R*)-2,6-Dimethylheptyl sulfate

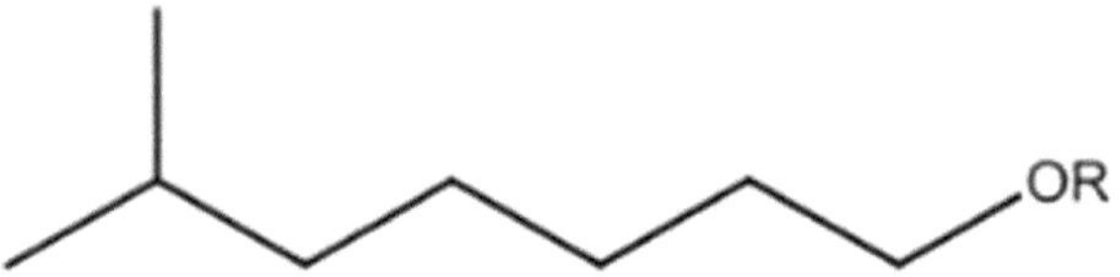

6-Methylheptyl sulfate (R = SO_3)

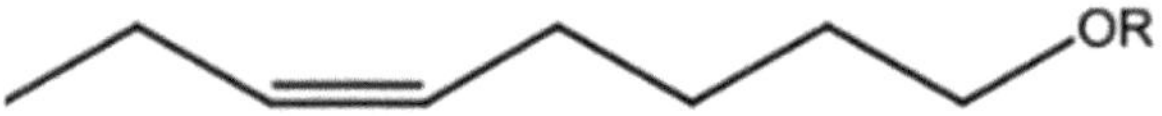

(*E*)-5-octenyl sulfate (R = SO_3)

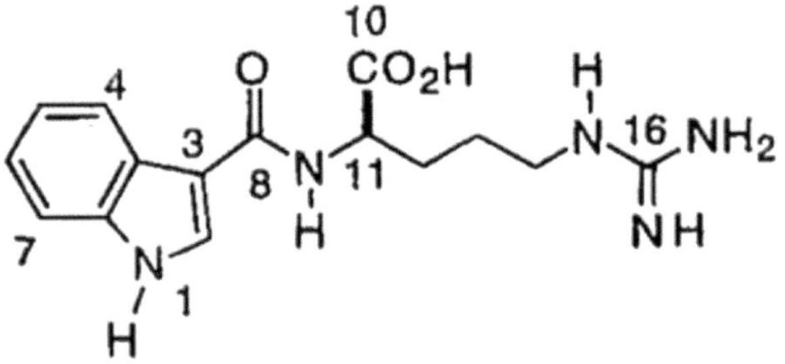

N-(1*H*-indolyl-3-carbonyl)-D-arginine

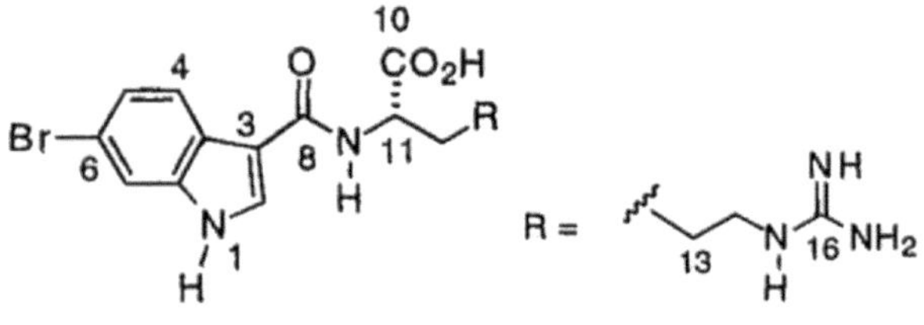

N-(6-bromo-1*H*-indolyl-3-carbonyl)-l-arginine

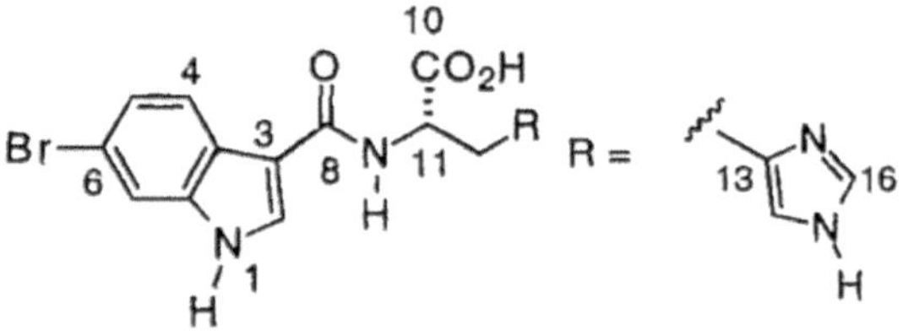

N-(6-bromo-1*H*-indolyl-3-carbonyl)-l-histidine

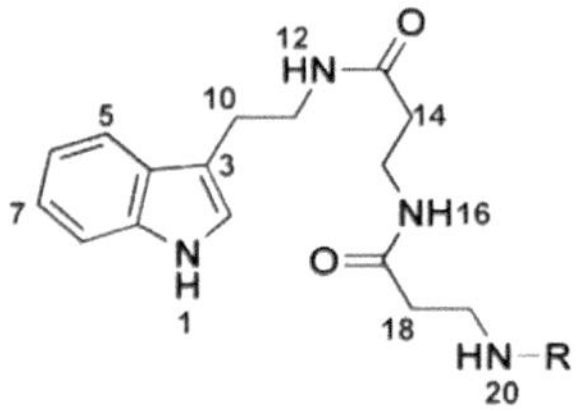

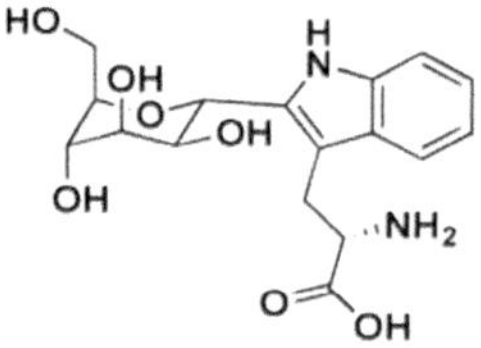

Leptoclinidamide C2-α-D-Mannosylpyranosyl-L-tryptophan

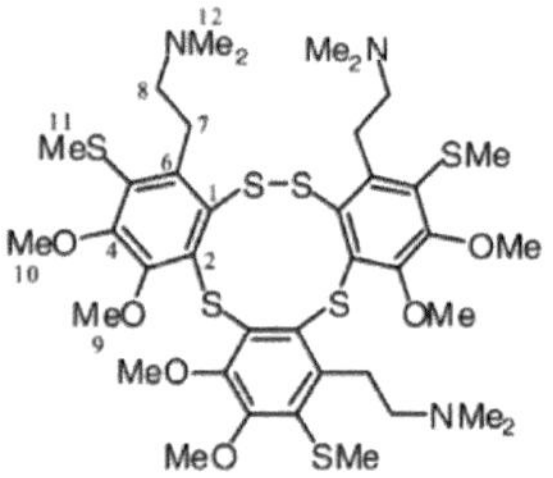

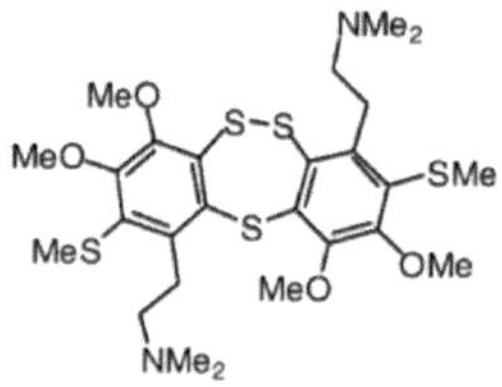

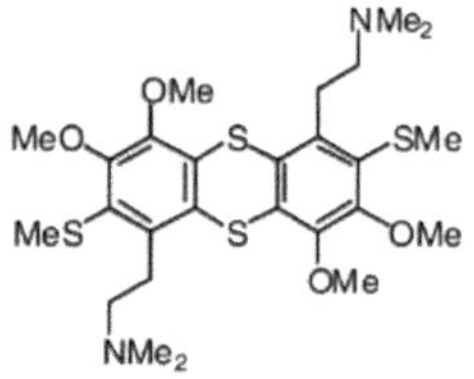

Lissoclibadin 1 Lissoclibadin 2 Lissoclibadin 3

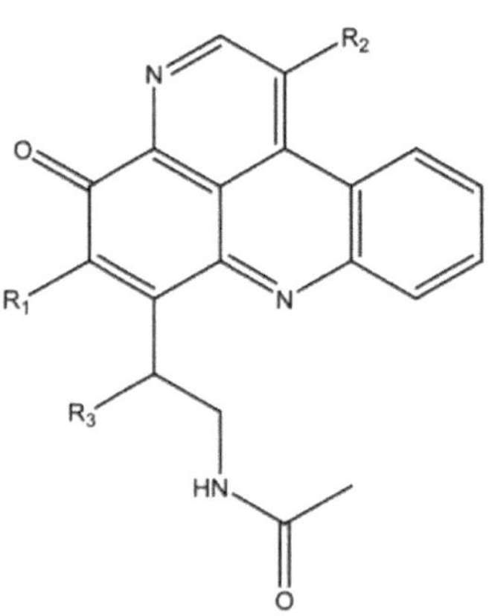

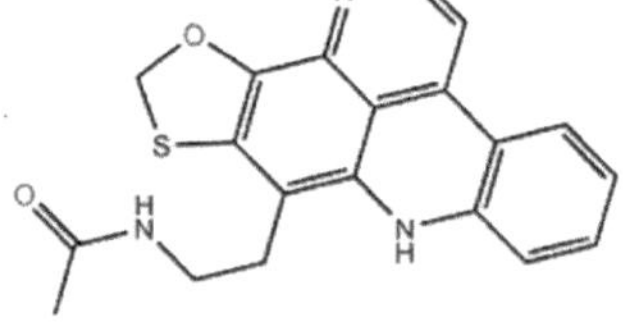

R1 = SCH_3, R2 = H, R3 = H

Diplamine Lissoclinidine

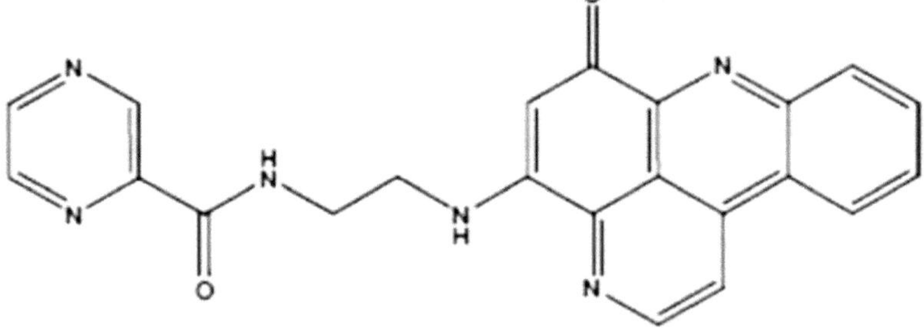

N-(2-(6-oxo-6*H*-pyrido[2,3,4-kl] acridin-4-ylamino)ethyl)pyrazine-2-carboxamide

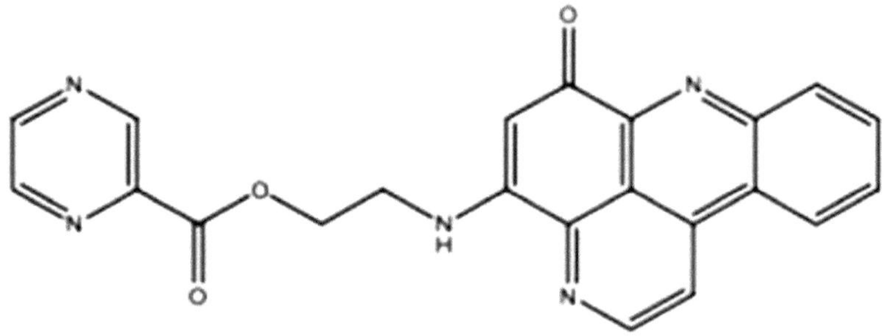

2-(6-oxo-6*H*-pyrido[2,3,4-kl]acridin-4-ylamino)ethyl pyrazine-2-carboxylate

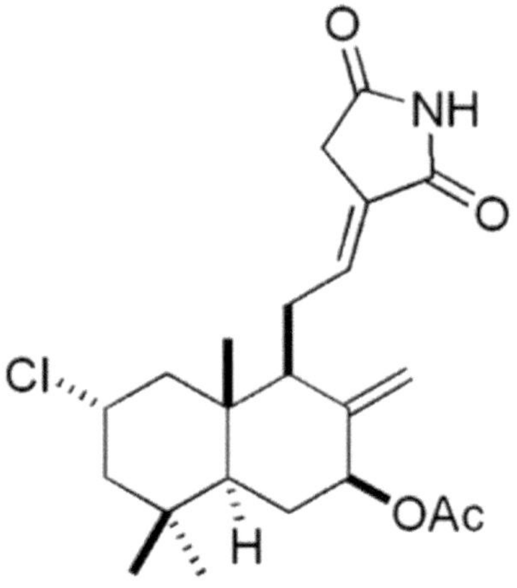

Haterumaimide O

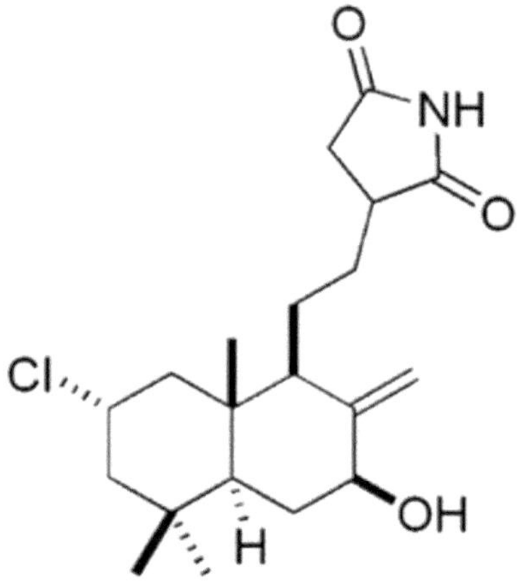

Haterumaimide P

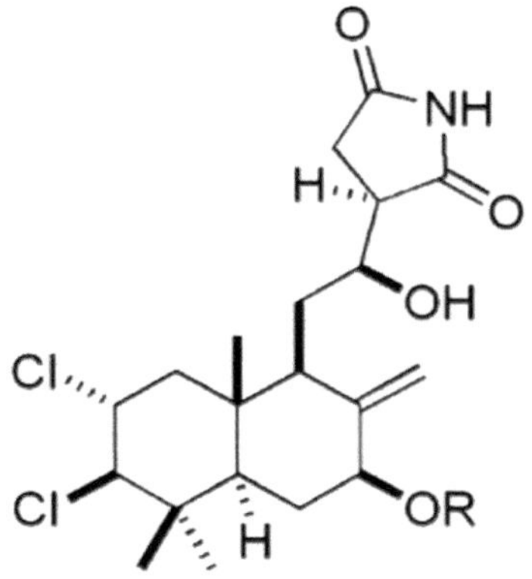

R = H

Dichlorolissoclimide

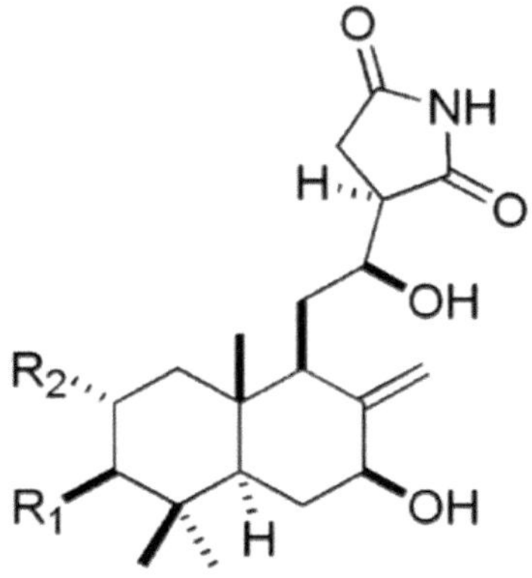

R1 = H, R2 = Cl

Chlorolissoclimide

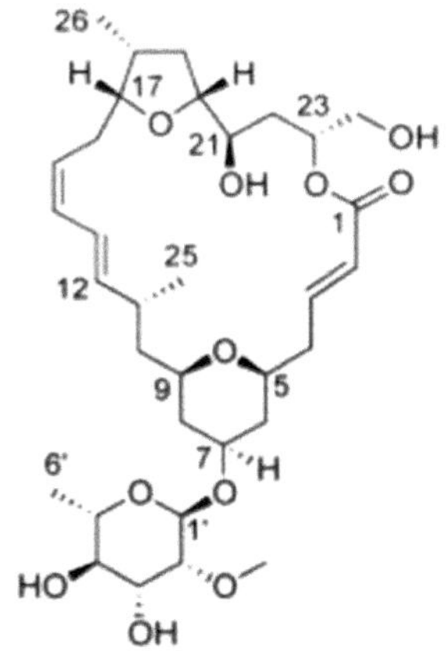

Mandelalide A

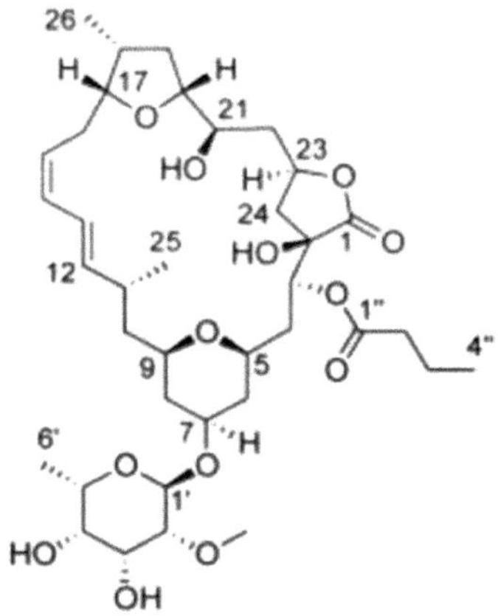

Mandelalide B

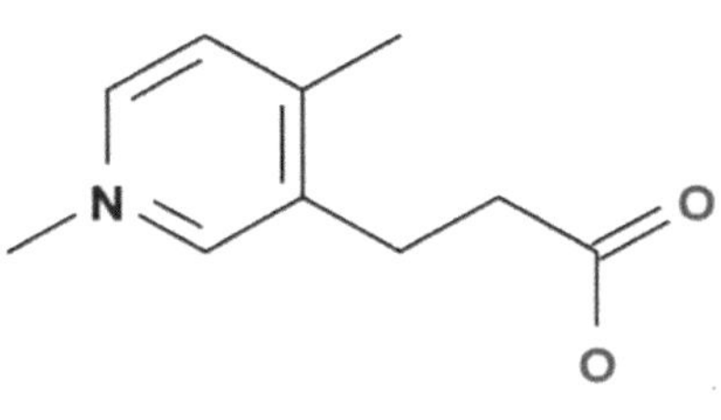

Sulcatin

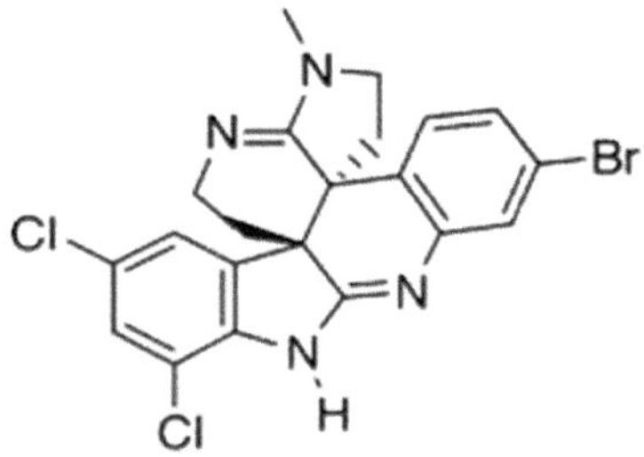

Perophoramidine

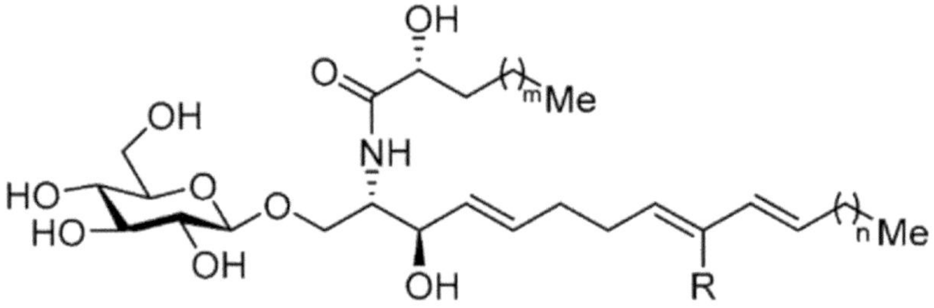

Phalluside 1 (R = CH_3, *m* = 12, *n* =6)
Phalluside 2 (R = Ch_3, *m* = 14, *n* = 6)

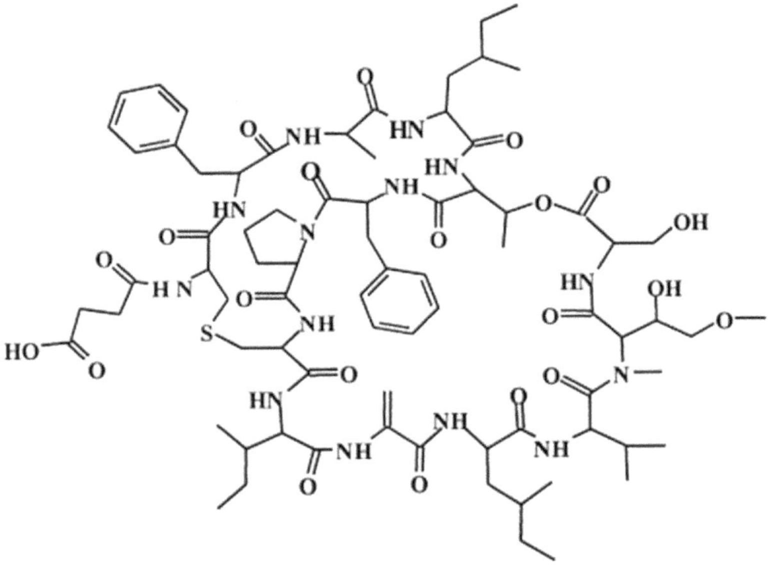

Vitilevuamide

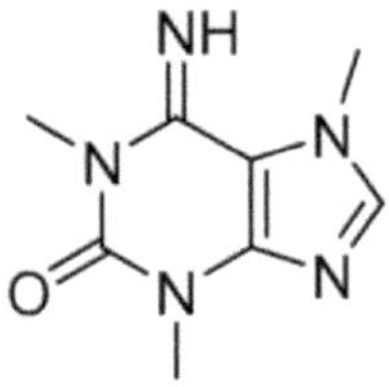

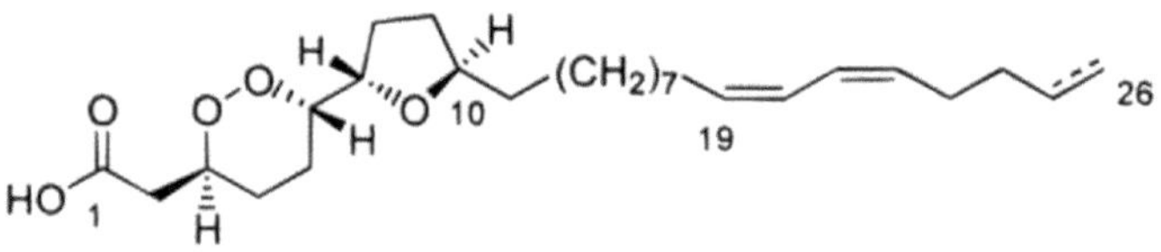

1,3,7-Trimethylisoguanine. Stolonic acid B; stolonic acid A (Δ^{25})

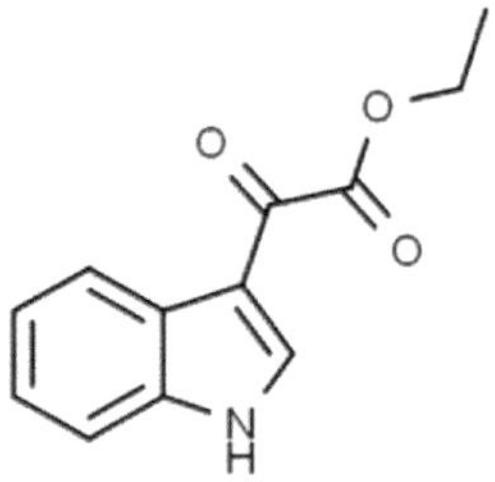

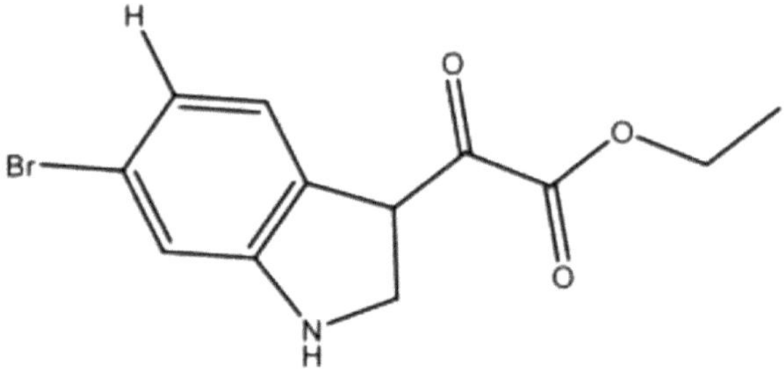

Ethyl indolyl-3-glyoxyiate Ethyl 6-bromoindolyl-3 glyoxylate

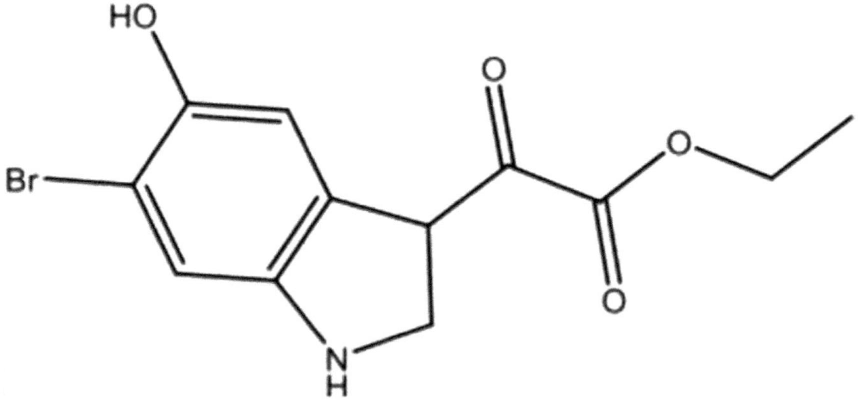

Ethyl 6-bromo-5-hydroxyindolyl-3-glyoxylate

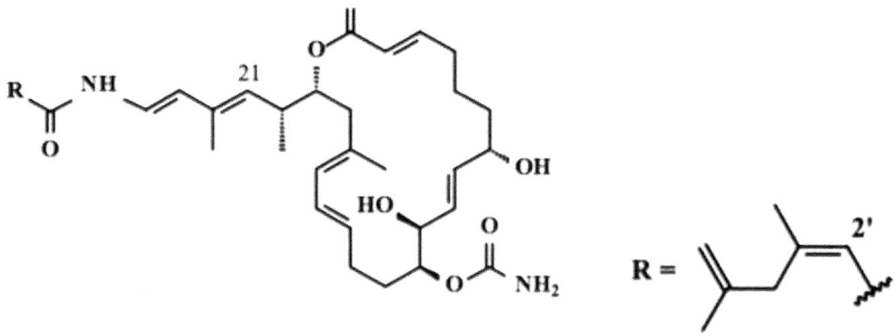

Palmerolide D

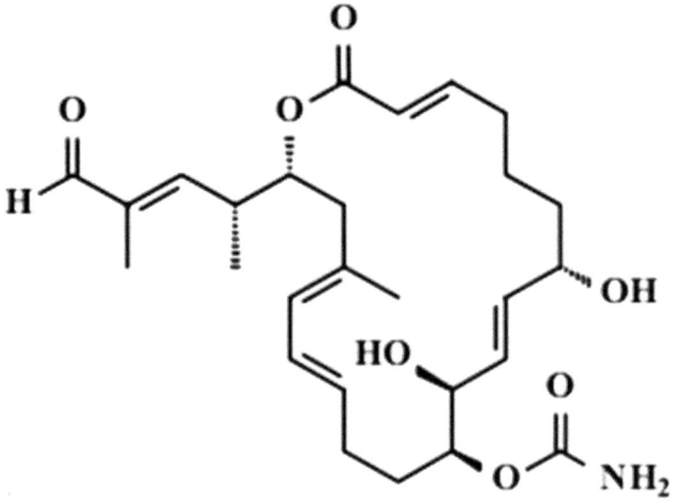

Palmerolide E

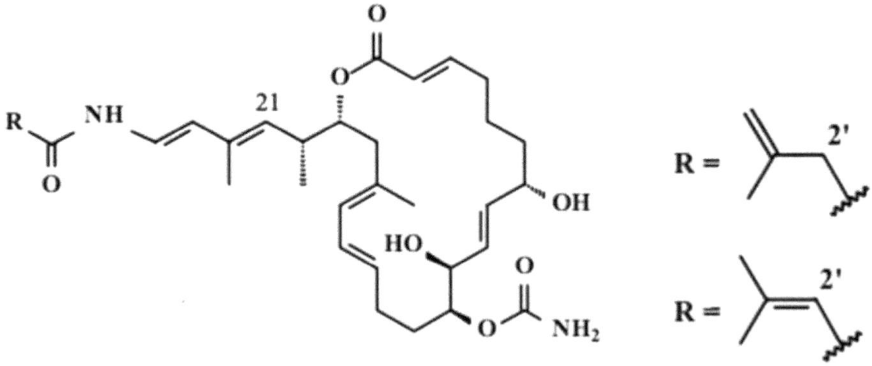

Palmerolide F, G

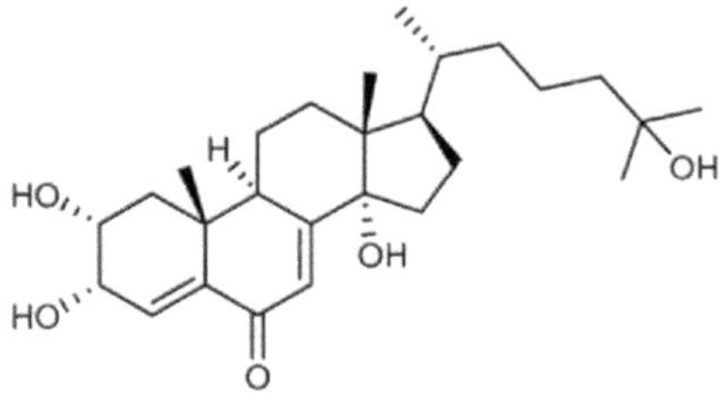

Hyousterone A

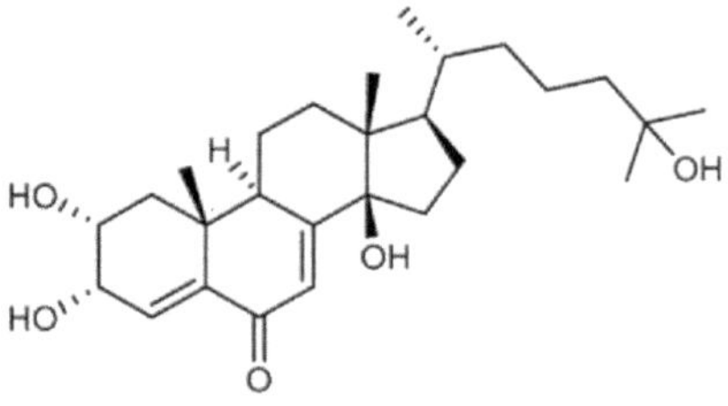

Hyousterone B

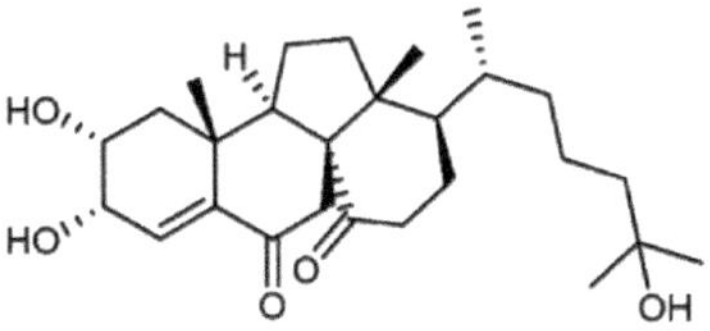

Abeohyousterone

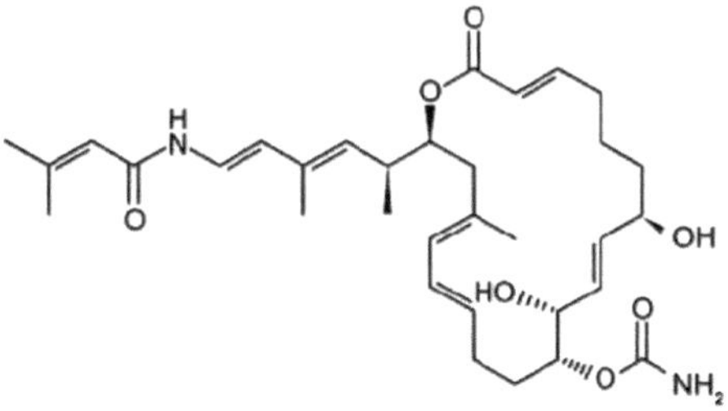

Palmerolide A

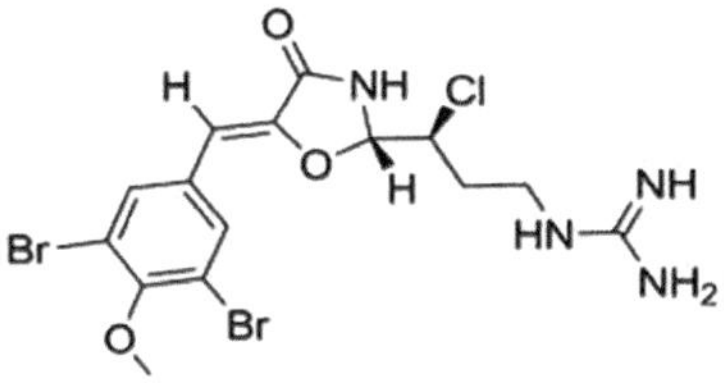

Synoxazolidinone A

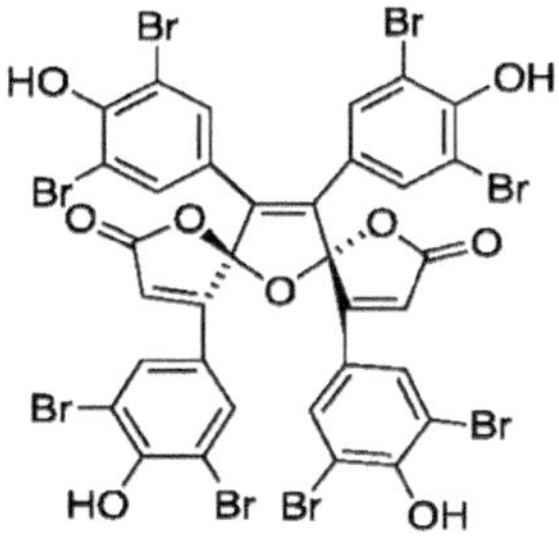

Prunolide A

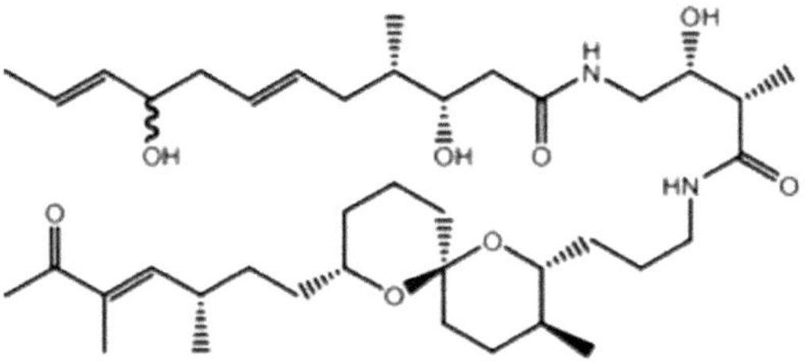

39-Oxobistramide K

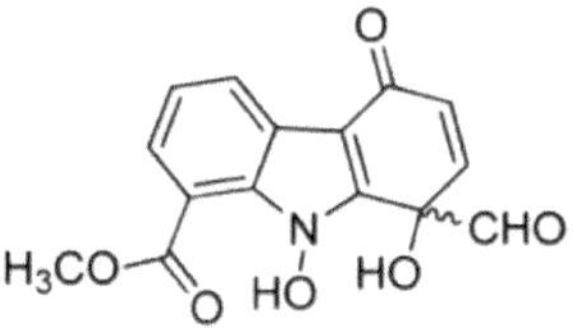

Coproverdine

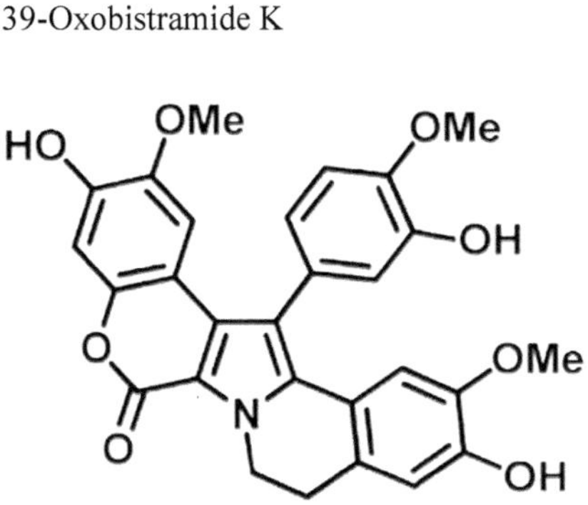

Lamellarin N

TABLE 6.5 Other Bioactivities of Ascidians (1)

Species	Compound(s)	Chemistry	Bioactivity	Mode of Action	Values
Aplidium albicans	*Plitidepsin* (generic name aplidine or dehydrodidemnin B; tradename Aplidin)	Cyclic depsipeptide	Anti-COVID-19	27.5 times more potent against SARS-CoV-2 than remdesivir	– (2)
Synoicum sp.	Cadiolide B; Prunolide A	Butenolide; tetraphenolic bis-spiroketal	Antiviral	Against potent RNA virus Japanese encephalitis virus	–
Didemnum molle	Homopolysaccharide	Sulfated mannose	Anti-HIV	Inhibit infection with CEM cells by HIV strain RF	–
-do-	Mollamides B	Hexapeptide	-do-	Against HIV-1 human PBM cells	EC50 48.7 μM
-do-	Mollamide F	-do-	-do-	-do-	39 and 78 μM
Unidentified sp.	Amellarin α 20-sulfate	Organic sodium salt	-do-	Against HIV-1 protease	IC50 16 μM
Didemnum sp.	Didemnidine B	Alkaloid	Antimalarial	Inhibition of *Plasmodium falciparum*	IC50 15 μM
Microcosmus helleri, *Ascidia sydneiensis*, and *Phallusia nigra*	Crude compounds	–	-do-	-do-	–
Aplidium meridianum	Meridianin G	Alkaloid	-do-	-do-	IC50 4.4 μM
Didemnum molle	Mollamide B	Hexapeptide	-do-	Against *Plasmodium falciparum* (D6 clone and W2 clone)	IC50 2.0, 2.1 μM
Didemnum albopunctatum	Albopunctatone	Anthrone-anthraquinone	-do-	Against chloroquine-resistant *Plasmodium falciparum*	IC50 5.3 μM

TABLE 6.5 *(Continued)*

Species	Compound(s)	Chemistry	Bioactivity	Mode of Action	Values
Didemnum sp.	Didemnidine B	Indole spermidine alkaloid	-do-	Inhibition against *Plasmodium falciparum*	IC50 8.4 μM
Aplidium scabellum	Scabellone B	Meroterpenoid	-do-	Against *Plasmodium falciparum* (K1 chloroquine-resistant strain)	IC50 4.8 μM
Aplidium sp.	Ascidiathiazone A	Dioxothiazino-quinoline-quinone	-do-	Against *Plasmodium falciparum* K1 strain	IC50 3.3 μM
Aplidiopsis confluata	Aplidiopsamine A	Polyaromatic alkaloid	-do-	Against chloroquine sensitive (3D7) strain of *Plasmodium falciparum*	IC50 = 1.47 μM
-do-	-do-	-do-	-do-	Against chloroquine resistant (Dd2) strain of *Plasmodium falciparum*	IC50 = 1.65 μM
Pseudodistoma opacum	Opacaline A, B	β-Carboline alkaloids	-do-	Against chloroquine-resistant strain of *Plasmodium falciparum*	2.5–14 μM (3)
Synoicum macroglossum	Tiruchanduramine	β-Carboline guanidine alkaloid	Antidiabetic activity	Moderate α-glucosidase inhibitory activity	IC50 78.2 μM
Didemnum obscurum	Lamellarin γ; Lamellarin K-triacetate	Lamellarin alkaloids	-do-	–	IC50 3.28, 2.96 μM
Halocynthia roretzi	Astaxanthin	Carotenoid pigment	-do-	Regulating lipid/ glucose metabolism	– (6)

TABLE 6.5 *(Continued)*

Species	Compound(s)	Chemistry	Bioactivity	Mode of Action	Values
Lissoclinum notti	*N*-(2-(6-oxo-6*H*-pyrido[2,3,4-kl] acridin-4-ylamino)ethyl) pyrazine-2-carboxamide; 2-(6-oxo-6*H*-pyrido[2,3,4-kl] acridin-4-ylamino)ethyl pyrazine-2-carboxylate	Alkaloids	Antituberculosis	Inhibition of Mycobacterium tuberculosis H37Rv	MIC 2 μM
Styela plicata and *Halocynthia pyriformis*	Dermatan sulfates	Sulfated glycosaminoglycans	Anticoagulant activity	Thrombin inhibition by heparin cofactor II	(4), (5)
Halocynthia roretzi	Crude extracts	–	Antiarthritic activity	Alleviated paw edema	(7)
Microcosmus squamiger	Methanol and dichloromethane extracts	–	Antioxidant activity	–	23.0 and 26.3 mg TE/g (8)
Ciona intestinalis	Relaxin	Polypeptide hormone	Sperm motility enhancing	–	(9)

Source: Adapted from (1) Palanisamy et al. (2017); (2) https://pharmacy.ucsf.edu/news/2021/02/cancer-drug-stops-sars-cov-2-lab; (3) Chan et al. (2011); (4) Pavão et al. (1998); (5) Gandra et al. (2000); (6) Zhu et al. (2022), (7) Hong et al. (2014); (8) Pan et al. (2021) and (9) Hassanzadeh (2011).

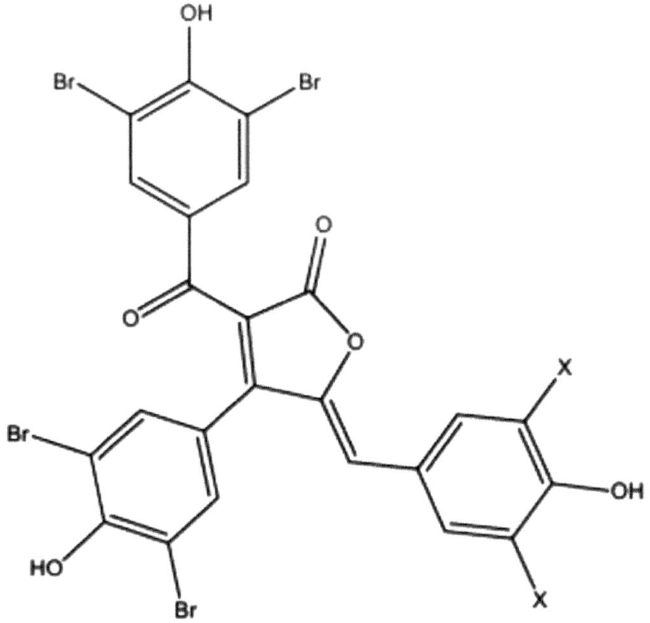

X = Br

Cadiolide B

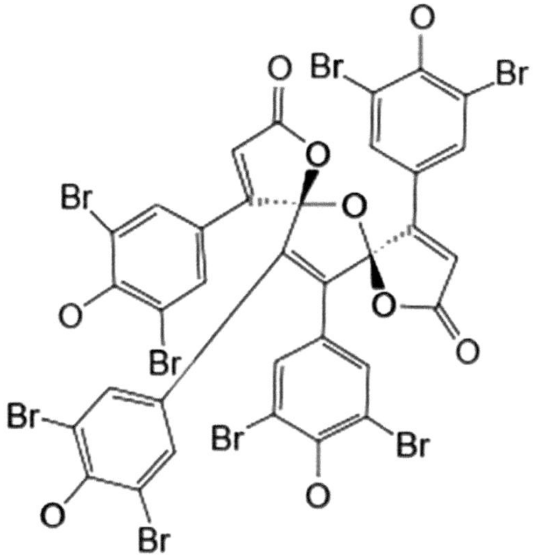

Prunolide A

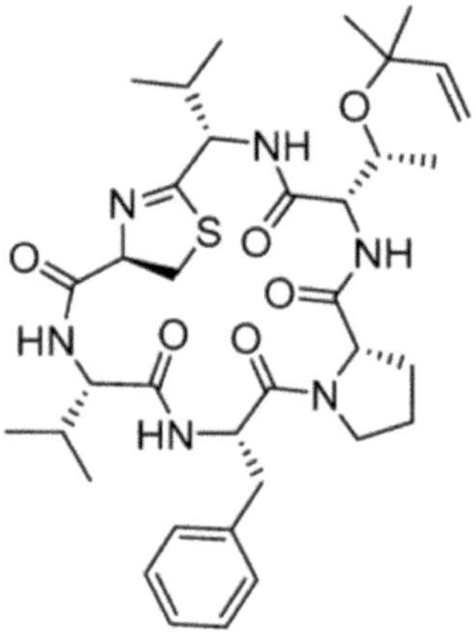

Mollamide B

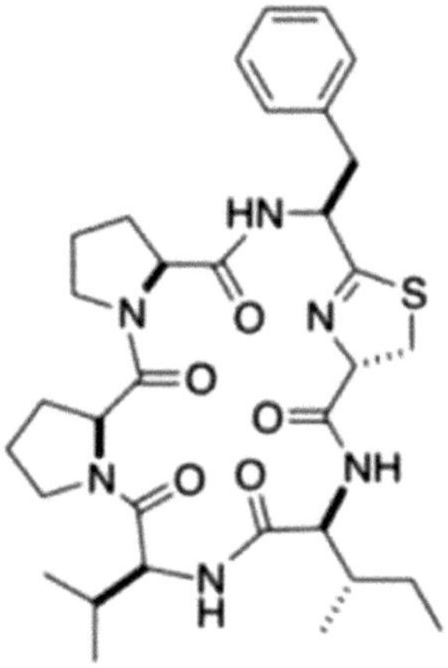

Mollamide F

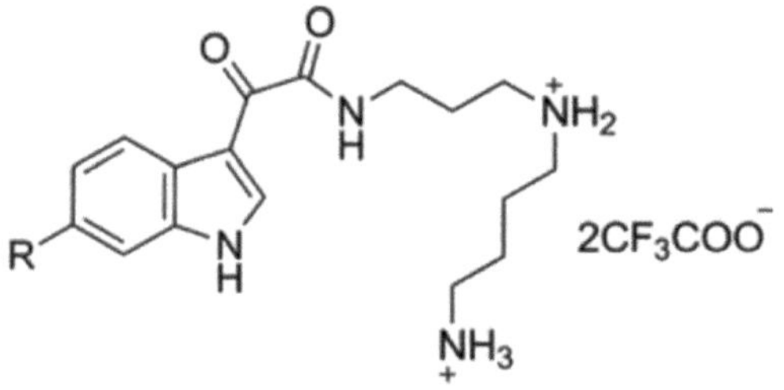

Didemnidine B

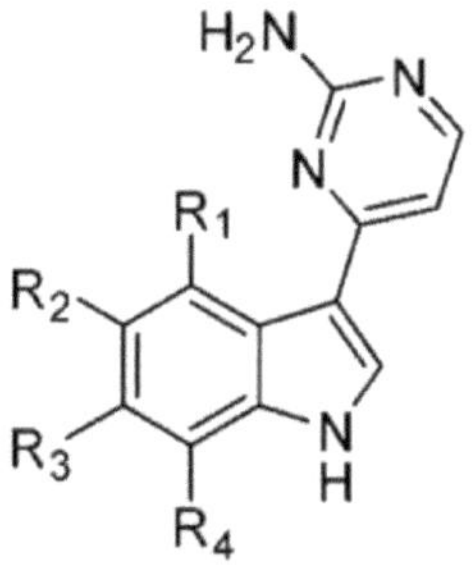

R1 = R2 = R3 = R4 = H

Meridianin G

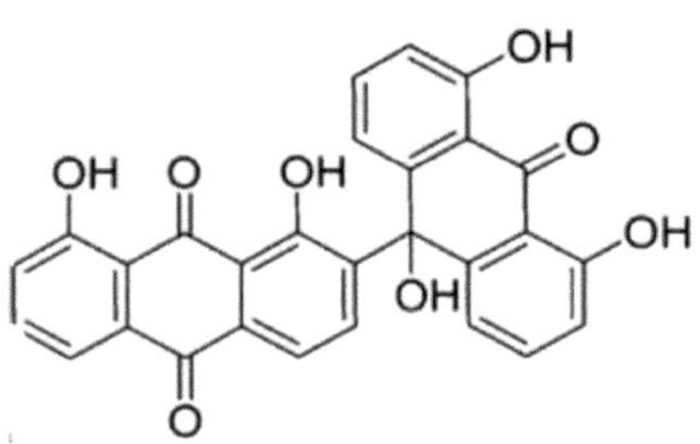

Albopunctatone

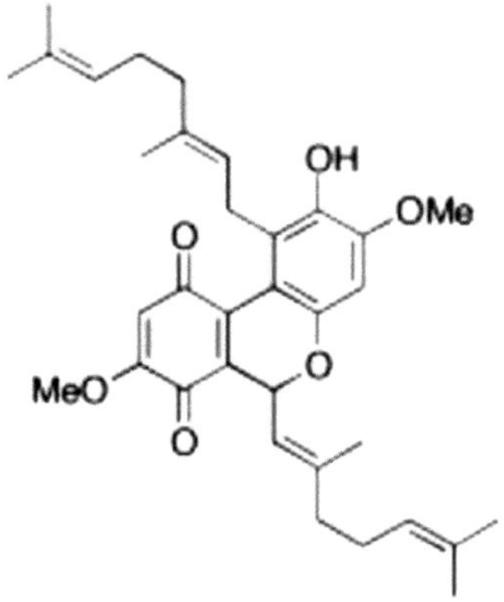

Scabellone B

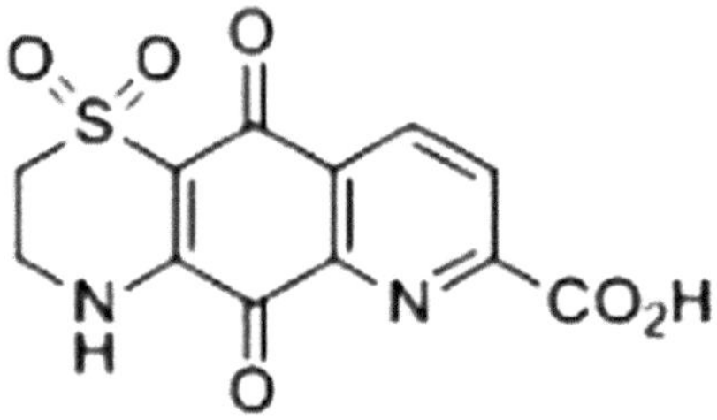

Ascidiathiazone A

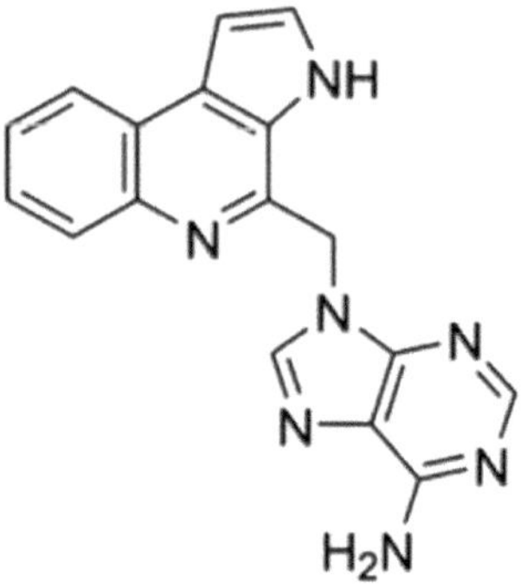

Aplidiopsamine A

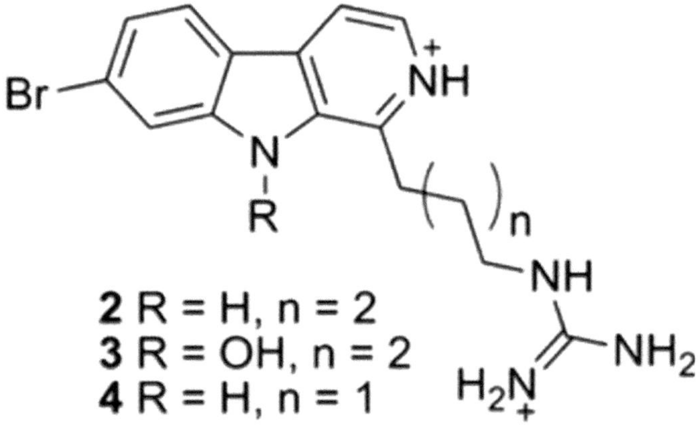

Opacaline A, B.

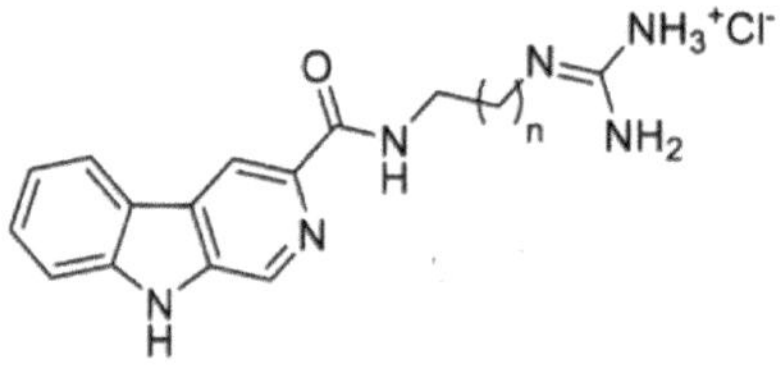

Tiruchanduramine

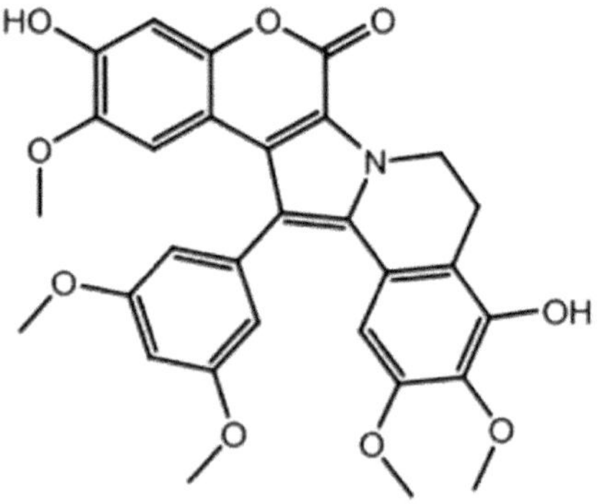

Lamellarin γ

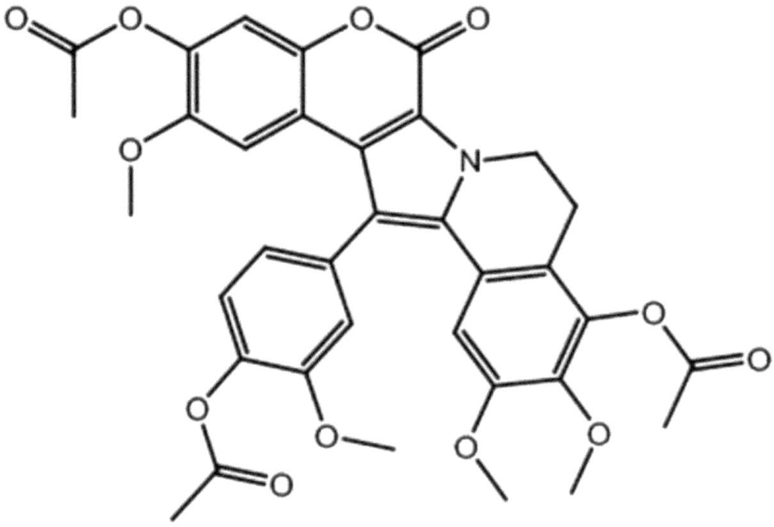

Lamellarin K-triacetate

6.5 BIOACTIVE POTENTIAL OF SYMBIOTIC ASSOCIATIONS IN ASCIDIANS

It is amazing to know that some of the secondary metabolites derived from the ascidians are truly produced by their associated (symbiotic) microbiome community consisting of bacteria, archaea, or fungi. Among the three ascidian orders, namely, Aplousobranchia, Phlebobranchia, and Stolidobranchia, the Aplousobranchia microbiomes have been the most studied (63%) followed by Phlebobranchia and Stolidobranchia (15% and 22%, respectively). Further, the ascidians of the families Didemnidae and Polyclinidae (Aplousobranchia); and Ascidiidae (Phlebobranchia) have been well studied and the most studied symbiotic ascidian genera include *Didemnum*, *Diplosoma*, *Lissoclinum*, *Aplidium*, *Ascidia*, and *Trididemnum*. It is also worth-mentioning here that the chemical classes' secondary metabolites produced by the symbiotic ascidian are alkaloids, terpenes, and polyketides which possess antibacterial, antitumoral, cytotoxic activities (Matos and Antunes, 2021). The different constituents of microbiome community associated with the ascidians and their secondary metabolites are given in Table 6.6.

TABLE 6.6 Microbiomes Associated with Ascidians and Their Bioactive Compounds

Ascidian	Associated Microbiome	Taxonomy of Microbiome	Bioactive Compound(s)	Chemical Class	References
Didemnum sp.	*Gordonia didemni*	Actinobacteria	–	–	de Menezes et al. (2016)
Ecteinascidia turbinata	*Salinispora arenicola*	-do-	Arenimycin	Benzo[*α*] naphthacene quinone class of antibiotics	Dou and Dong (2019)
Polysyncraton lithostrotum	*Salinispora pacifica*	-do-	Lomaiviticin	Antibiotic	Matos and Antunes (2021)
Eudistoma toealensis	*Salinispora and Verrucosispora*	-do-	–	–	Matos and Antunes (2021)
Trididemnum orbiculatum	*Nocardia* strain	-do-	Peptidolipins	Lipopeptides	Matos and Antunes (2021)
Ascidia ahodori, *Ascidia sydneiensis samea*, *Ciona intestinalis*, *Eudistoma toealensis*, *Styela plicata*, and *Synoicum adareanum*	–	Flavobacteria (bacteroidetes)	–	–	Matos and Antunes (2021)
Lissoclinum patella	–	Bacteroidetes	–	–	Behrendt et al. (2011)
Lissoclinum patella	–	Fusobacteriota (fusobacteria)	–	–	Behrendt et al. (2011)
Ascidia ahodori, *Ascidia sydneiensis samea*, and *Styela plicata*	–	Planctomycetes	–	–	Matos and Antunes (2021)
Didemnum molle	*Nitratireductor basalti-strain UMTGB225*	Alphaproteobacteria (proteobacteria)	–	–	Matos and Antunes (2021)

TABLE 6.6 *(Continued)*

Ascidian	Associated Microbiome	Taxonomy of Microbiome	Bioactive Compound(s)	Chemical Class	References
Synoicum adareanum	*Pseudovibrio* sp.	Alphaproteobacteria (proteobacteria)	α-*N*-Acetylgalacto-saminidase	–	Matos and Antunes (2021)
Lissoclinum patella	*Candidatus Endolissoclinum faulkneri*	Alphaproteobacteria (proteobacteria)	Patellazoles A–C	Polyketides	Matos and Antunes (2021), Dou and Dong (2019)
Ecteinascidia turbinata	*Agrobacterium* sp.	Alphaproteobacteria (proteobacteria)	Sesbanimide A	Alkaloid	Dou and Dong (2019)
Polycitonidae sp.	*Agrobacterium* sp.	Alphaproteobacteria (proteobacteria)	Sesbanimide C	Alkaloid	Dou and Dong (2019)
Aplidium conicum, *Ascidiella scabra*, *Botryllus schlosseri*, *Cystodytes dellechiajei*, *Diazona violacea*, and *Pseudodistoma cyrnusense*	*Roseobacter* sp.	Alphaproteobacteria (proteobacteria)	–	–	Matos and Antunes (2021)
Tridideminum solidum	*Tistrella mobilis and Tistrella bauzanensis*	Alphaproteobacteria (proteobacteria)	Didemnin B	Cyclic depsipeptide	Dou and Dong (2019)
Ascidia, *Ascidiella*, *Botryllus*, *Ciona*, *Didemnum*, *Eudistoma*, and *Styela*	*Endozoicomonas* sp.	Gammaproteobacteria (proteobacteria)	–	–	Matos and Antunes (2021)
Ciona intestinalis	*Pseudoalteromonas tunicata*	Gammaproteobacteria (proteobacteria)	–	–	Matos and Antunes (2021)
Stomozoa roseola (=Stomozoa murrayi)	*Acinetobacter* sp.	Gammaproteobacteria (proteobacteria)	Bromoindole-3-carbaldehyde	Indole derivative	Matos and Antunes (2021)

TABLE 6.6 *(Continued)*

Ascidian	Associated Microbiome	Taxonomy of Microbiome	Bioactive Compound(s)	Chemical Class	References
Ecteinascidia turbinata	*Candidatus Endoecteinasci frumentensis*	Gammaproteobacteria (proteobacteria)	ET-743 (Yondelis®)	FDA-approved drug	Matos and Antunes (2021)
Synoicum adareanum	*Microbulbifer* sp.	Gammaproteobacteria (proteobacteria)	Palmerolide A	Macrolide	Matos and Antunes (2021)
Lissoclinum patella	–	Thaumarchaeota, Euryarchaeota, and Crenarchaeota (archaea)	–	–	Matos and Antunes (2021)
Cystodytes dellechiajei, *Distaplia bermudensis*, *Eudistoma toealensis*, *Herdmania momus*, *Polyandrocarpa anguinea*, *Polyandrocarpa zorritensis*, and *Styela plicata*	*Nitrosopumilus* sp.	Thaumarchaeota (archaea)	–	–	Matos and Antunes (2021)
Didemnum sp.	–	Filamentous fungi	–	–	Matos and Antunes (2021)
Didemnum molle	*Trichoderma virens*	Ascomycota (fungi)	Trichodermamide A, B	Modified dipeptides	Matos and Antunes (2021)
Didemnun ligulum	*Trichoderma harzianum*	Ascomycota (fungi)	–	–	Matos and Antunes (2021)
Unidentified ascidian	*Penicillium citrinum* and *Fusarium proliferatum*	Ascomycota (fungi)	–	–	Matos and Antunes (2021)

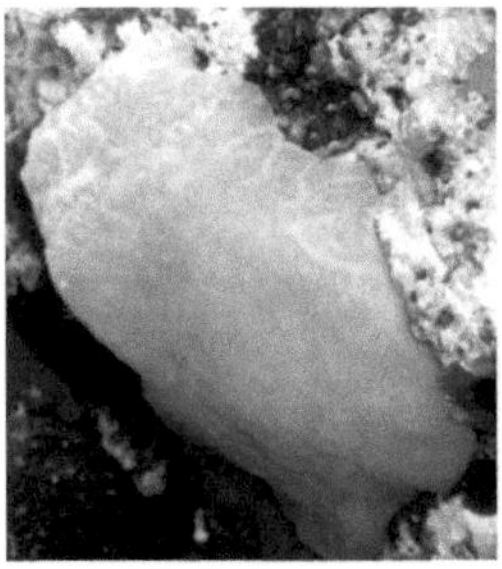

1.

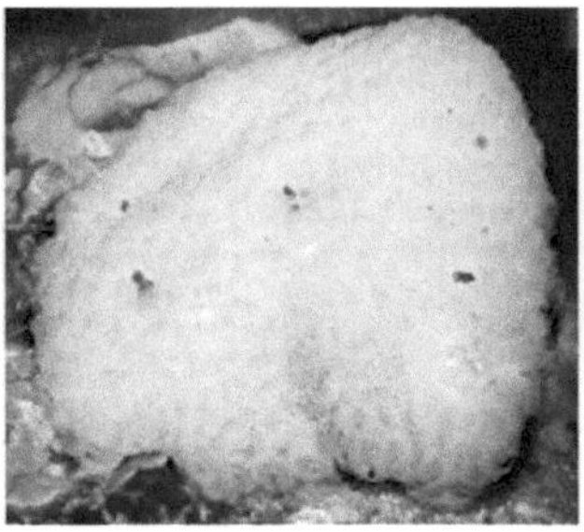

2.

3.

4.

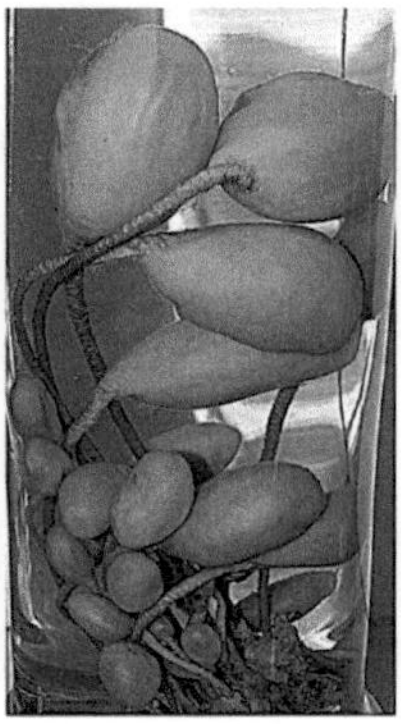

5.

6.

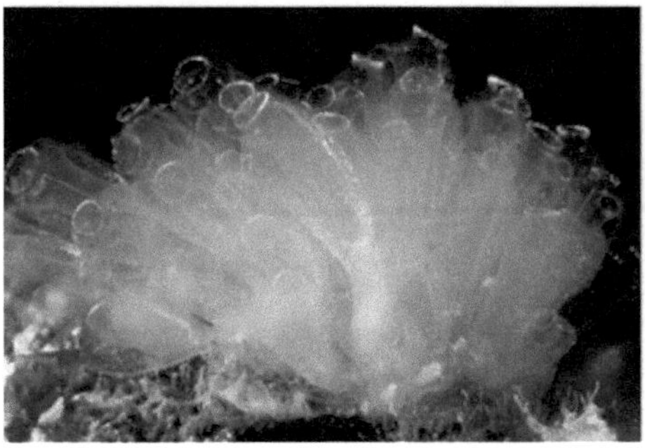

7.

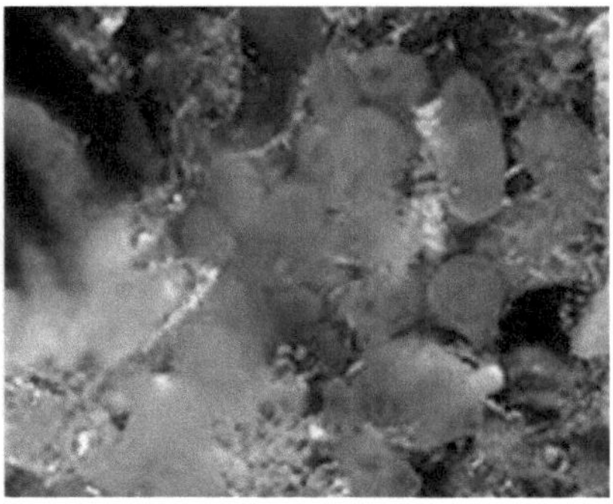

8.

9.

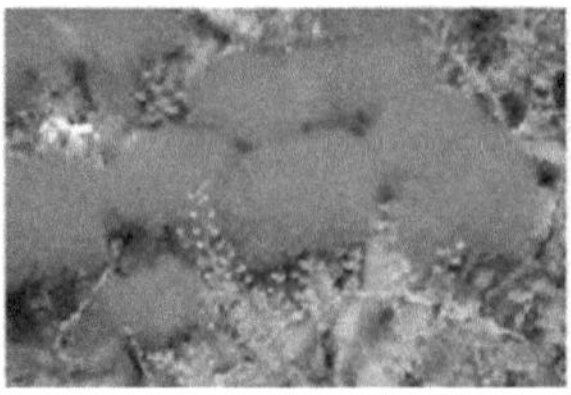

10.

11.

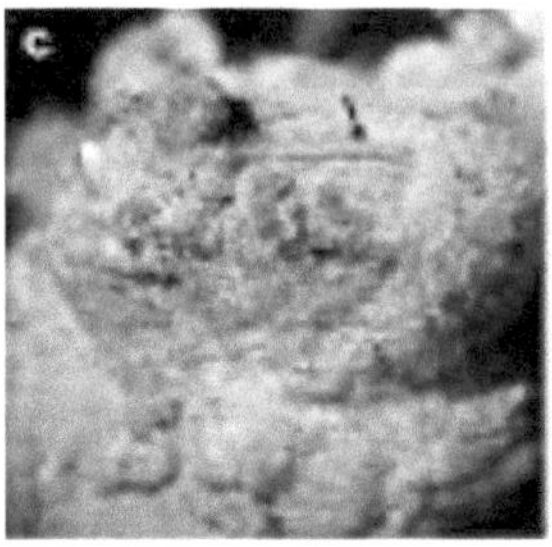

12.

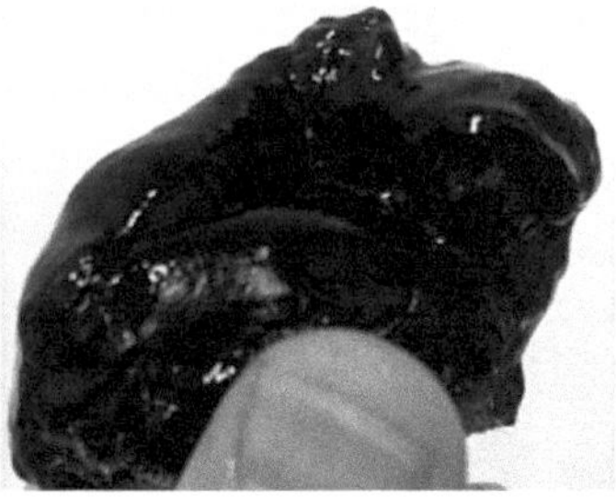

13.

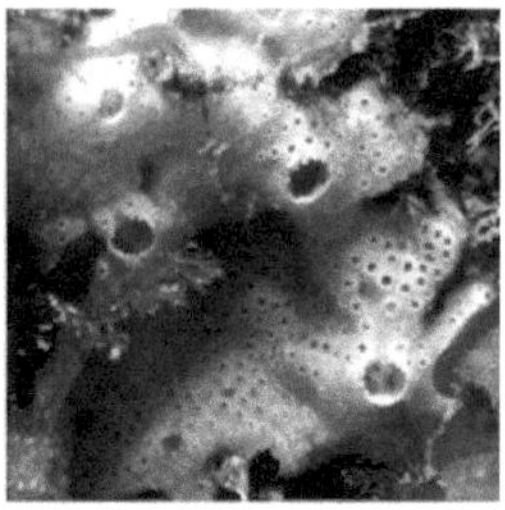

14.

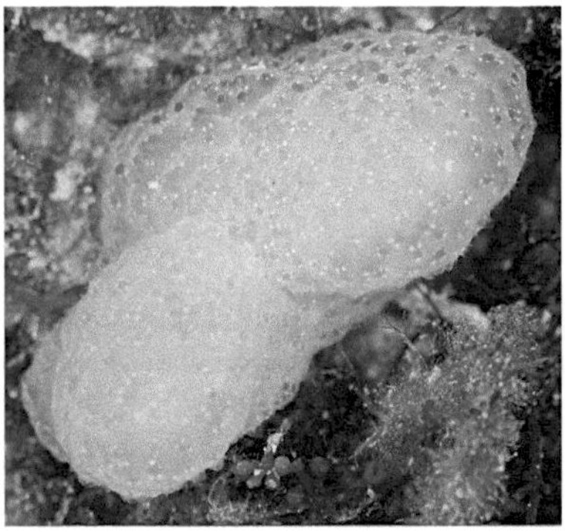

15.

16.

1. *Aplidium conicum*
   Image credit: Elapied. https://creativecommons.org/licenses/by-sa/3.0
2. *Aplidinium falklandicum*
   Image credit: Reprinted with permission from Dirk Schories
3. *Aplidium meridianum*
   Image credit: TreatmentBank, Public domain
4. *Ascidia mentula*
   Image credit: Matthieu Sontag. https://creativecommons.org/licenses/by-sa/3.0/

5. *Boltenia ovifera*
   Image credit: A. C. Tatarinov -. https://creativecommons.org/publicdomain/zero/1.0/
6. *Ciona savignyi*
   Image credit: Steve Lonhart / NOAA MBNMS. Public domain.
7. *Clavelina phlegraea*
   Image credit: Reprinted with permission from Mario munaretto.
8. *Diazona angulata*
   Image credit: Plazi TreatmnetBank. Public domain.
9. *Didemnum mole*
   Image credit: Univ of Guelph students. https://creativecommons.org/licenses/by/3.0/
10. *Eudistoma vannamei*
    Image credit: Reprinted with permission from Paulay, Gustav.
11. *Halocynthia papillosa*
    Image credit: Elapied. https://creativecommons.org/licenses/by-sa/2.0/fr/
12. *Microcosmus helleri*
    Image credit: Plazi. Public domain.
13. *Phallusia nigra*
    Image credit. Jorge Stolfi. https://creativecommons.org/licenses/by-sa/3.0/
14. *Polysyncranton* sp. (=*Polysyncranton lithostrotum*) Image credit: Géry PARENT. Public domain.
15. *Pseudodistoma* sp. (*Pseudodistoma cereum*)
    Image credit: Diego Delso. https://creativecommons.org/licenses/by-sa/4.0/
16. *Trididemnum cyclops*
    Image credit: Philippe Bourjon. https://creativecommons.org/licenses/by-sa/4.0/

CHAPTER 7

ASCIDIAN CELLULOSE IN BIOMEDICAL APPLICATIONS

Cellulose is a natural polymer and is composed of repeating glucose units $(C_6H_{10}O_5)_n$. It has been reported to be the most excellent polysaccharide and organic material on earth. Based on the sources of origin, cellulose is categorized into bacteria-based cellulose, algae-based cellulose, plant-based cellulose, wood-based cellulose, and ascidian-based cellulose. The cellulosic particles take different morphological forms, namely, cellulose crystals, cellulose filaments, cellulose fibers, and cellulose micro/nano-fibrils. In this chapter, the biomedical applications of the ascidian-based cellulose are dealt with.

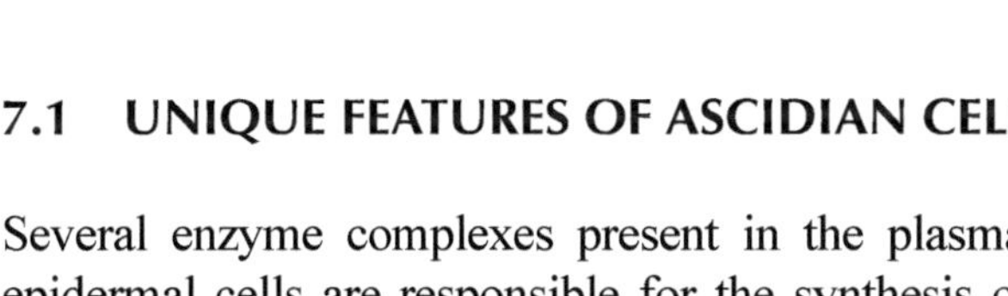

Cellulose

7.1 UNIQUE FEATURES OF ASCIDIAN CELLULOSE

Several enzyme complexes present in the plasma membrane of ascidian epidermal cells are responsible for the synthesis of cellulose. Hundreds of cellulose nanofibrils are found bundled in the tunic which is the outer tissue of ascidian. From this tunic, a pure form of cellulose called "tunicin" is extracted. The purified extracted cellulose from tunic is known as ascidian cellulose or tunicin. The shape and dimensions of a nanofibril bundle of the tunic may vary

Biotechnological Potential of Ascidians: Wacky and Weird Sea Squirts. Ramasamy Santhanam (Author)

depending on the species and the nanofibril bundles get deposited in a multi-layered texture parallel to the surface of the epidermis. The ascidian cellulose or tunicin is nothing but the extracted and purified cellulose from tunic. The dimensions and shape of a nanofibril bundle derived from the tunic may differ based on the species of ascidian and nature of deposition in a multilayered texture parallel to the epidermis. The length of nanofibrils produced by the ascidian cellulose ranges from 100 nm to >2 μm) and width from 10 nm to 30 nm. Among the different tunics, the dry tunic has been reported to contain 27% nitrogen-containing components and about 60% cellulose. However, after extraction, the ascidian cellulose is highly crystalline (ca. 95%) and is composed of nearly pure cellulose (Seddiqi et al., 2021). The typical dimensions of the two types of ascidian nanocellulose, namely, cellulose nanofibril and cellulose nanocrystal, are shown in Table 7.1.

TABLE 7.1 Dimensions of Types of Nanocellulose Derived from Ascidians

Type of Nanocellulose	Length (nm)	Width (nm)
Cellulose nanofibril	>2000	10–30
Cellulose nanocrystal	100–2000	5–30

Source: Adapted from Seddiqi et al. (2021).

The cellulose whiskers produced by the ascidians have been reported to possess exciting applications. These whiskers have the width of around 15 nm, and lengths between 1160 and 2000 nm. The high levels of crystallinity (95%), specific surface area, high aspect ratio of these cellulose whiskers due to hydroxyl groups provide them with high amount of stiffness, mechanical properties, and composite reinforcement.

The cellulose nanocrystals are derived from the cellulose fibers by milling the fibers first, followed by alkali and bleaching treatment; and hydrolysis or mechanical homogenization. The methods of preparing cellulose from the tunic of ascidians as reported by Hassanzadeh (2011) are given below.

7.2 PRODUCTION OF ASCIDIAN CELLULOSE

7.2.1 PREPARATION OF CELLULOSE

The starting material for the production of cellulose is the root removed ascidian sample which is dried first and is reduced to powder by using a knife mill. The size of ascidian sample powder used in the pulping process is less than 40 meshes. The ratio of all treating solutions: pulp is 10:1. A 5-g ascidian tunic powder is first dipped in an acidic aqueous solution of H_2SO_4 (0.9 wt%) kept in a stainless steel vessel and is heated at 180°C with gentle stirring. Then the ascidian tunic is filtered by using a glass filter, washed with acetone/water, and dried under vacuum at 75°C. In the next step, the acid-treated sample is treated with alkali aqueous solution of $NaOH/Na_2S$ (9/3 wt%) for 2 h at 180°C and is filtered, washed, and dried as in the previous step. Now the alkali-treated sample is treated with aqueous NaOCl solution (2.9 wt%) as a bleaching agent for 1 h at 75°C, and then it is also filtered, washed, and dried as in the previous processes. The final yield is then calculated and carbohydrate analysis is done for measuring the cellulose content.

7.2.2 FERMENTATION OF ASCIDIAN SAMPLES

For the fermentation of the ascidian samples, 0.25 g ascidian sample is added to a 250-mL beaker containing 72% sulfuric acid. This solution is autoclaved and is then neutralized with solid barium hydroxide ($Ba(OH)_2$). For about 30-mL sample, 5 g $Ba(OH)_2$ is added and the solution is centrifuged and is divided into two portions. The first part of the solution which is about 200 mL is kept under vacuum evaporation. When the solution reaches near one-fifth of its first volume, 1 mL is taken from this stock solution and is diluted for measuring the sugar. The rest of the solution is considered for fermentation analysis as two sets which are treated with two strains of yeast. The ethanol produced by both strains of yeasts is measured once every 24 h (Hassanzadeh, 2011).

7.3 ASCIDIAN CELLULOSE IN BIOMEDICAL APPLICATIONS

As compared to synthetic biopolymers, the cellulose of ascidians possesses unique features such as biodegradability, biocompatibility, nontoxicity, excellent mechanical properties, low production cost, and so on. These features make the ascidian cellulose to play a significant role in biomedical applications including drug delivery, tissue engineering, and wound dressing.

7.3.1 TISSUE ENGINEERING

The tissue engineering as a medical practice is utilizing biomaterials to create constructs which are exposed to tissues and organs of the human body. For tissue engineering, cellulose of ascidians could serve as a valuable scaffold material or additive as it has mechanical properties matching real tissues.

The cellulose nanofibrils (ACN) derived from the ascidian, *C. intestinalis* could be a boon for cartilage and soft-tissue repair when they are combined with living cells. The main advantages of ACN are similar dimensions to collagen, cost-efficient extraction procedures, and feasible industrial production. Further, the mechanical properties of ascidian nanocellulose are found to be superior to nanocellulose extracted from agricultural sources, and the ACN's biocompatibility is also very much comparable to that of ePTFE (expanded polytetrafluoroethylene), an exemplary biomaterial. Furthermore, the ascidian ACN is the most safe and suitable biomaterial with amazing rheological properties for applications with 3D-bioprinting technology. Recent research findings have also reported that ACN has its great advantage for use in tissue engineering as it can be autoclaved or sterilized by radiation or electron beam (Apelgren et al., 2022; Seddiqi et al., 2021). Hydroxyethyl cellulose, hydroxypropyl cellulose, cellulose acetate, cellulose, methyl cellulose, carboxymethyl cellulose (CMC), and ethyl cellulose are the most commonly used cellulose derivatives in tissue engineering. Similar derivatives from ascidian nanocellulose offer vast scope in the filed of tissue engineering.

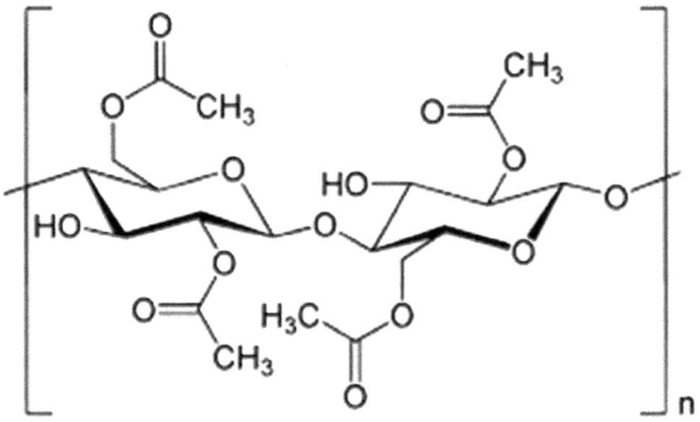

Cellulose acetate

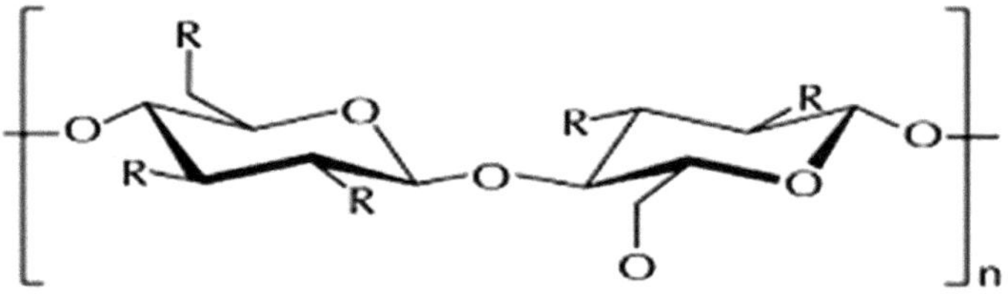

Hydroxyethyl cellulose

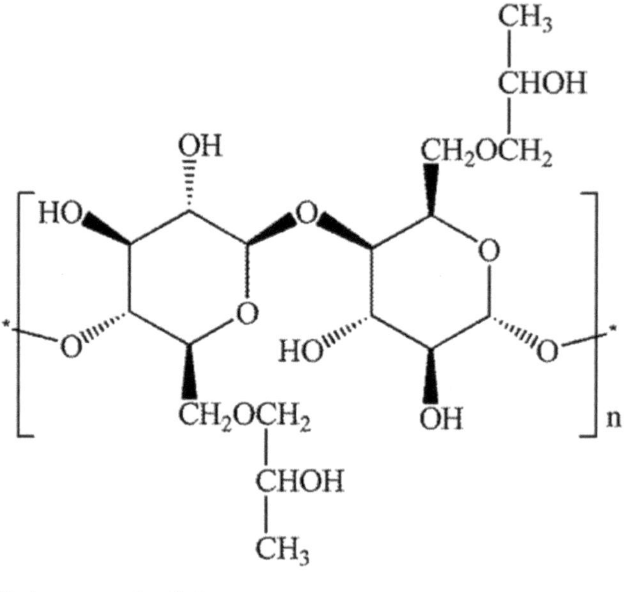

Hydroxypropyl cellulose

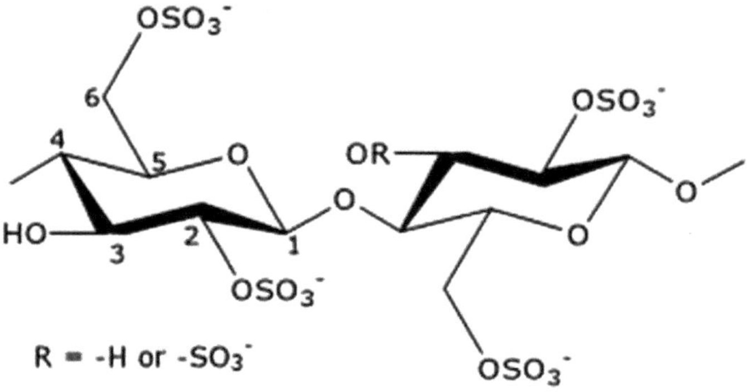

Cellulose sulfate

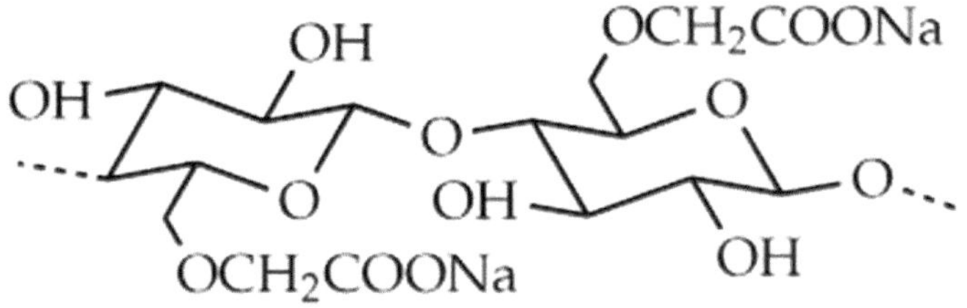

Carboxymethyl cellulose

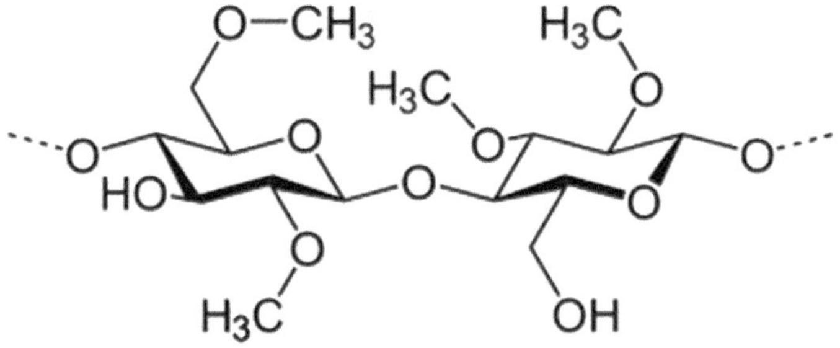

Methyl cellulose

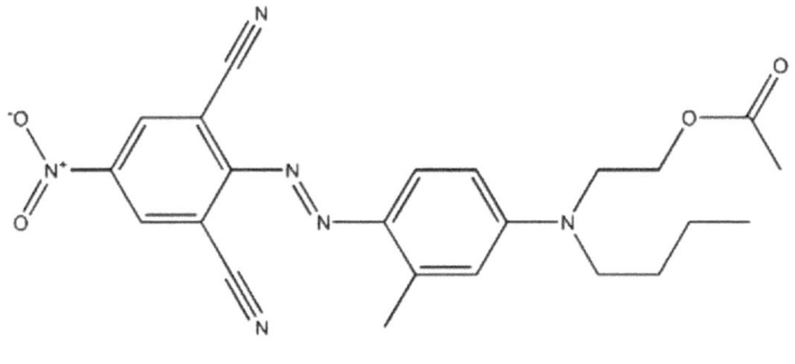

Ethyl cellulose

7.3.2 WOUND DRESSING

Cellulose-based are known to serve as novel wound dressing materials for burn wounds and skin grafting. As wound dressing materials, these polymers promote high water absorption capacity, water retention, porosity, and dryness to help avoiding avoids bacterial attachment. Normally, the cellulosic material may be shaped like a drug-loaded sheet and is used as a cover or bandage on wounds. The nanocellulose derived from ascidians has shown amazing results in wound healing practices, and this is largely due to its water retention property and moisture absorption. When this nanocellulose is applied over the wound, it may not only lower the inflammatory responses but would also promote the proliferation of fibroblasts in the wound healing cascade. Further, the nanofibrillar cellulose of ascidians could also be a suitable matrix for wound healing due to its high porosity, high water-holding ability, and high surface area-to-volume ratio. Further, ascidian-based nanofibrillar cellulose hydrogel could also be a novel substance for mitigating surplus wound contraction in vivo and in vitro. Furthermore, the nanofibrillar cellulosic, namely, CMC has been reported to be the most promising wound dressing cellulosic material which is commonly used in the treatment of ulcers and burns due to its high water-bonding affinity, swelling capacity, biocompatibility with sin and mucous membrane, biodegradability, nontoxicity, low immunogenicity, and so on. Dissolvable CMC foam may also be employed as wound dressing, especially after sinus surgery. In these ventures, the application of CMC helps in the formation of synechia and lowering of postoperative bleeding. Oxidized cellulose nanofiber (CNF) has also been reported to be the most suitable material for wound healing applications because of its well-dispersed cellulose fibrils and adequate water absorption capacity (Seddiqi et al., 2021).

7.3.3 DRUG DELIVERY

Cellulose is commonly used in the pharmaceutical industry in orally administered tablets in which I serves as a coating blended with several excipients. However, the biomedical applications of cellulose as drug carriers for extended drug release is yet to be implemented and this is largely due to their uncertainty in promoting controllable diffusive properties and dissolvability in drug delivery system. In this regard, it is worth mentioning here that cellulose has been reported to demonstrate amazing drug-delivery patterns through instant and controlled release, especially in oral dosages. Further, cellulosic materials have been reported to be more suitable for use as enteric coatings on capsules or tablets due to their resistance to the acidic conditions of the stomach. However, ascidian-derived CNFs offer much scope in biomedicinal applcations as a suitable carrier for controlled drug delivery due to their favorable flexibility and elasticity, low toxicity, and low density that can be employed for grafting specific groups. Further, ascidian-derived cellulose acetate-based drug-loaded nanofibers may also have wide attention in the future, especially in the discovery of transdermal and topical drug delivery systems owing to their advantageous characteristics, such as biodegradability, biocompatibility, high shear strength and shear modulus, high affinity with other substances, regenerative properties, and amicable flexural and tensile strength. Further, cellulose acetate phthalate electrospun fibers have been reported to help resisting HIV infections. These fibers have also been also found very suitable for preventing HIV infection and loading anti-HIV drugs during sexual intercourse (Seddiqi et al., 2021).

CHAPTER 8

ASCIDIANS AS A SOURCE OF INDUSTRIALLY IMPORTANT BYPRODUCTS

Several species of ascidians are known to possess byproducts like collagens, sulfated polysaccharides, glycosaminoglycans, sterols, and trace elements which show a wide range of functionalities and are considered to be potential materials for pharmaceutical, nutraceutical, and chemical industries (Ryu and Kim, 2014). The species of ascidians possessing these byproducts and their importance are dealt with in this chapter.

8.1 COLLAGENS

Collagens, a family of extracellular matrix proteins, are known to yield structure and support to the human skin, muscles, bones, and connective tissues. Information is available on the molecular species of collagen in ascidian tissues is scanty. The collagen present in the body wall of the ascidian *C. intestinalis* and collagen from the muscle layer of the food ascidian, *H. roretzi* have been reported (Mizuta et al., 2002).

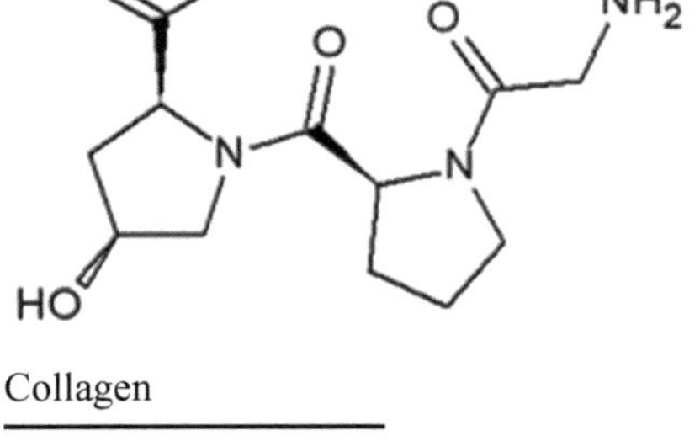

Collagen

Biotechnological Potential of Ascidians: Wacky and Weird Sea Squirts. Ramasamy Santhanam (Author)

8.2 POLYSACCHARIDES

Sulfated polysaccharides are a complex group of macromolecules possessing a wide range of important biological functions. While the epidermal cells of the tunic synthesize large amounts of sulfated polysaccharides such as sulfated L-galactan, dermatan sulfate-like glycosaminoglycan is seen in their body. Dermatan sulfates are present in ascidians such as *Phallusia nigra, Halocynthia pyriformis, S. plicata*, H. pallida, *H. roretzi and C. intestinalis* (Kozlowski et al., 2011; Pavão, 1996).

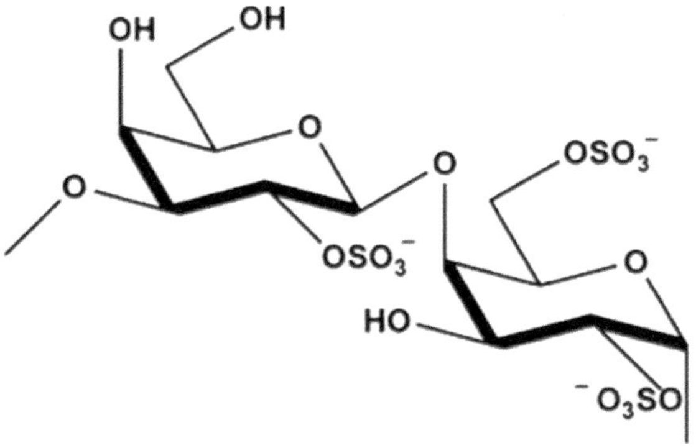

Sulfated polysaccharide

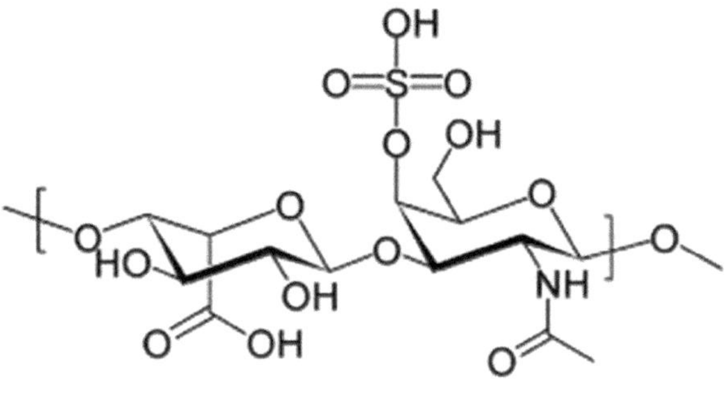

Dermatan sulfate

8.2.1 SULFATED GLYCOSAMINOGLYCANS

These are oversulfated dermatan sulfates or mucopolysaccharides seen in several species of ascidians such as *S. plicata, H. pyriformis, P. nigra*, and

C. intestinalis. Glycosaminoglycans are seen in the extracellular matrix of several organs and are consisting of a linear sequence of repeating disaccharide units formed by an amino sugar (*N*-acetylglucosamine and *N*-acetylgalactosamine) and an uronic acid (glucuronic acid or iduronic acid) or galactose. These compounds are known to play significant roles in human physiological activities such as cell proliferation, cell division, cell–cell and cell–matrix adhesion, morphogenesis, and the regulation of signaling molecules (Karamanou et al., 2017).

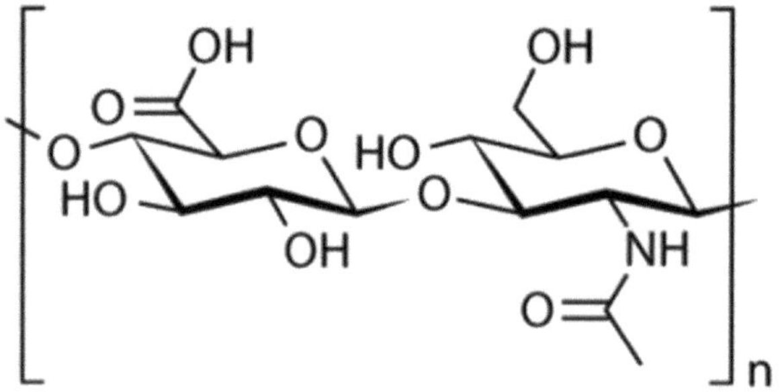

Sulfated glycosaminoglycans

8.2.2 *HEPARAN SULFATE PROTEOGLYCANS*

These low sulfated glycans are glycoproteins and are seen in the ascidian, *S. plicata*. These compounds possess the common characteristic of containing one or more covalently attached heparan sulfate (HS) chains.

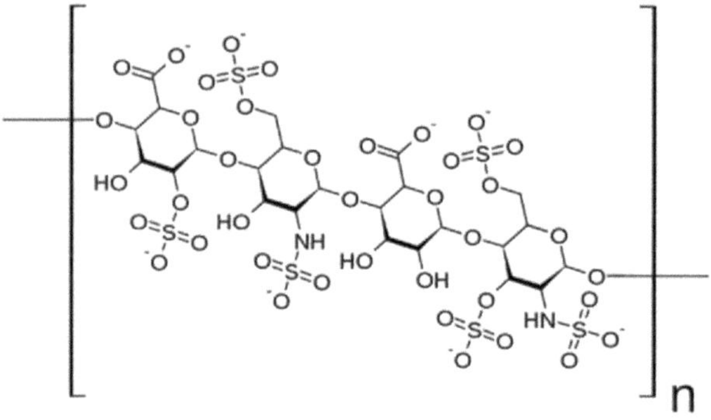

Heparan sulfate

8.3 STEROLS

The ascidians, *Microcosmus sulcatus* and *Halocynthia papillosa*, have been reported to possess small amounts of C30 sterols (Voogt and Rheenen, 1975). A paper by Yasuda in 1975 reported that sterols of *S. plicata* and *S. clava* are a complex mixture of stanol and Δ5-sterols. In these species, cholesterol was the dominant sterol and cholestanol was also in considerable amount. In *Polycitor* sp., a total of 27 previously known sterols have been reported and these sterols consisted of C26–C29 steroid alcohols (Makar'eva et al., 1995). C26 derivatives amounted to 12.4% of the total steroid fraction; and C27, C28, and C29 contributed 49.74%, 27.95%, and 4.4%, respectively. Further, there was only 33% of sterols with saturated side-chains, and the amount of cholesterol in the mixture was 9.53%. The sterol profile of this species is shown in Table 8.1.

TABLE 8.1 Sterol Profile of *Policitor* sp.

Sterol	Percentage
5alpha~-cholestan-3beta-ol	18.7
24-Methylcholesta-5,22-dien-3beta-ol	14.8
Cholest-5-en-3beta-ol	9.5
5 alpha~-Cholest-22-en-3beta-ol	7.3
24-Nor-5a-cholest-22-en-3beta-ol	6.2
24-Methyl-5alpha~-cholest-24(28)-en-3beta-ol	5.9
24-Methyl-5alpha-cholest-22-en-3beta-ol	5.32
24-Norcholesta-5,22-dien-3beta-ol	5.31
Choleta-_5,22-dien-3beta-ol	4.2
24-Methylcholesta-5,24(28)-dien-3beta-ol	3.5
5alpha~-Cholesta-7,22-dien-3beta-ol	3.1
7-Nor-24-methylcholesta-5,22-dien-3beta-ol	2.7
27 -Nor-5alpha--methylcholest-22-en-3beta-ol	1.95
5alpha~-Cholesta-7,22-dien-3beta-ol	1.80
4-Ethylcholest-5 -en-3beta-ol	1.76
24-Ethylcholesta-5,22-dien-3beta-ol	1.2
24-Methyl-5alpha~-cholesta-7,22-dien-3beta-ol	1.1
24-Nor-5 a-cholesta-7,22-dien-3beta-ol	0.93
24-Methyl-5alpah~-cholesta-7,24(28)-dien-3beta-ol	0.87
24-Ethyl-5alpha~-cholest-22-en-3beta-ol	0.78
27-Nor-5alpha-24-methylcholesta-7,22-dien-3beta-ol	0.45

TABLE 8.1 *(Continued)*

Sterol	Percentage
24-Ethyl-5alpha~-cholest-22-en-3beta--ol	0.39
24-Ethyl-5alpha~-cholest-7-en-3beta-ol	0.15
4-Ethyl-5alpaha~-cholesta-7,22-dien-3beta-ol	0.11
24-Methylcholest-5-en-3beta-ol	Trace
24-Methyl-5alpha~-cholestan-3beta-ol	Trace
24-Methyl-5alpha-cholest-7-en-3beta-ol	Trace

Source: Adapted from Makar'eva et al. (1995)

Kijajic et al. (1983) reported on the sterol profile of *M. sulcatus* and *H. papillosa* and both the species showed more or less equal pattern of sterol composition as shown in Table 8.2.

TABLE 8.2 Sterols of *Microcosmus sulcatus* and *Halocynthia papillosa*

Steol Type	Sterol% in Microcosmus sulcatus	Sterol% in Halocynthia papillosa
Cholesterol	16.5–33.4	40.1–42.0
Cholestanol	52	–
Cholest 7-en-3beta-ol	15	–
Fucosterol	4.2–6.7	0.5–3.5
C30 sterols	0.3–1.2	0.2–0.5
Occelasterol	4.9–6.1	4.4–5.2
24-Ethyl-cholest-22-en-3beta-ol	1.0	–
24-Ethyl-cholest-5-en-3beta-ol	8.0–10.2	8.7–12.5
24-Methyl-cholest-5-en-3beta-ol	3.2	3.2–4.0
24-Ethyl-cholesta-5,22-dien-3beta-ol	5.4–11.8	4.2–4.3
(22*E*)-24-norcholesta-5,22-dien-3beta-ol	1.0–2.1	0.7–1.5
(22*E*)-cholesta-5,22-dien-3beta-ol	4.5–9.7	5.8–7.9
Brassicasterol	16.2–26.8	16.6–21.3
Cholesta-5.7-dien-3beta-ol + 24-methyl-cholest-22-en-3beta-ol	26.0	–
24-Methylene-cholest-5-en-3beta-ol + 24-methylene-cholestenol + 24-methylene-cholesta-5,7-dien-3beta-ol	5.6–8.1	6.4–7.2

Source: Adapted from Kijajic et al. (1983).

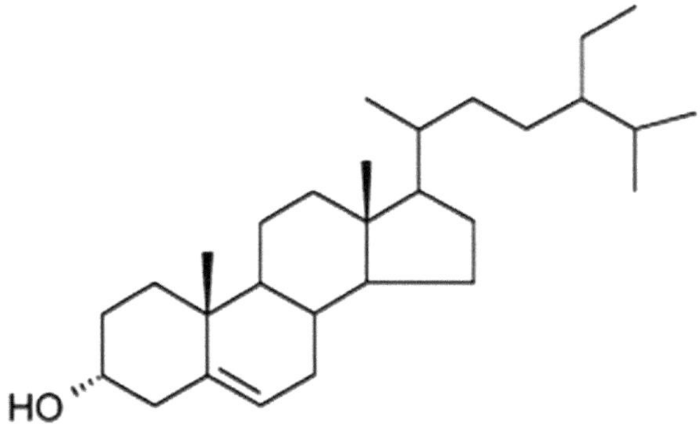

Sterol

8.4 TRACE ELEMENTS

Nutritionally important trace elements include iron, copper, cobalt, zinc, iodine, molybdenum, selenium, chromium, and vanadium.

8.4.1 IMPORTANCE OF TRACE ELEMENTS IN HUMAN PHYSIOLOGY (Rivas, https://www.osmosis.org/answers/trace-elements)

Iron: It has an important role in transporting oxygen throughout the human body through the blood. Iron deficiency may lead to anemia (deficiency of healthy red blood cells) and this deficiency has also been linked to the upper alimentary tract cancers.

Copper: It is believed to be the third most abundant trace element in the human body, and it is known to work with iron to keep healthy red blood cells. Further, it is also an essential component of many enzymes involved in the physiological reactions of the body. Furthermore, it also plays a significant role in maintaining the health of blood vessels, bones, and nerves.

Cobalt: It is found both in organic and inorganic forms. In the organic form, it forms an important part of vitamin B12 (AKA cobalamin) besides contributing to the formation of amino acids. On the other hand, the inorganic forms of cobalt is said to be toxic to the human body.

Zinc: It is significantly associated with the cell growth, cell division, tissue repair, and metabolic functions. It is also helping the immune system by inhibiting the entry of viruses and bacteria into the human body organs.

Iodine: It is a very important element present in the thyroid hormones, namely, triiodothyronine (T3) and thyroxine (T4). These hormones are also found to be very important for the growth, metabolism, and development of the human body.

Molybdenum: It is essential for certain enzymatic functions involved in the digestion and excretion in the human body.

Selenium: It is known to play an important metabolic role as an antioxidant activity which prevents or reduces damage caused by oxidation in the body.

Chromium: It plays a key role in regulating sugar, protein, and fat levels in the blood.

Vanadium: Vanadium compounds have been reported to induce potent hypoglycemic effects in animals with type 1 and type 2 diabetes mellitus. Jackson in 2000 also reported that vanadium possesses antidiabetic properties, possibly through its antihypertensive effects.

Though the ascidians are known to possess a variety of trace elements, reports are available on the iron, chromium and vanadium content of certain species of ascidians and are presented in Tables 8.3–8.5.

TABLE 8.3 Iron Content of the Tunic and Mantle of Ascidians

Species	Tunic (ng/mg dw)	Mantle (ng/mg dw)
Ascidia malaca	1684	494
Ascidia abhodori	357	564
Ascidia sydneiensis	1433	277
Ciona intstinalis	370	148
Ciona savignyi	88	55
Chelyosoma siboja	284	349
Halocynthia roretzi	51	15
Halocynthia aurantium	83	76
Halocynthia papillosa	44	61
Microcosmus sulcatus	72	68
Molgula manhattensis	7588	155
Phallusia mamillata	118	127
Pyura sacciformis	319	99
Polycarpa cryptocarpa	1710	80
Styelea plicata	289	43

Source: Adapted from Michibata et al. (1986).

TABLE 8.4 Chromium Content of Ascidians

Species	Ppm Dry Weight (max)
Aplidium californicum (=Amaroucium californicus)	7.95
Aplidium solidum (=Amaroucium solidum)	8.91
Ascidia ceratodes	5.94
Botrylloides sp.	8.4
Ciona intstinalis	12.1
Clavelina huntsman	16.5
Cystodytes lobaus	3.55
Distalpia occidentalis	5.99
Distalpia smithi	2.8
Eudistoma molle	15
Eudistoa psammion	152.5
Eudistoma ritteri	25.6
Perophora annectens	7.65
Policlinum planum	7.05
Ritterella pulchra	69.6
Ritterella rubra	45.2
Styela montereyensis	1.98
Synoicum parfustis	83.6

Source: Adapted from https://searchworks.stanford.edu/view/10773219

TABLE 8.5 Vanadium Content in the Mantle and Tunic Tissues of Ascidians

Species	Mantle (mM)	Tunic (mM)
Ascidia ahodori	11.2	2.4
Ascidia sydneinsis	0.7	0.06
Halocynthia rotetzi	0.001	0.01
Halocynthia aurantium	0.002	0.002
Ciona intestinalis	0.7	0.003
Phallusia mamillata	0.9	0.03
Styela plicata	0.001	0.005

Source: Adapted from Michibata (1996).

8.5 PRODUCTION OF PROTEASE BY ASCIDIANS AND ITS APPLICATION ON BLOOD STAIN REMOVAL

The marine ascidian, *Phallusia mammillata*-associated bacterium, *Virgibacillus halodenitrificans* RSK CAS1 has been reported to aid in the production of the enzyme protease in the presence of shellfish waste, namely, shrimp shell powder. The enzyme so produced would serve as a cleaning additive in blood stain removal thereby its possible use in the laundry detergent industry. In the blood stain removal experiments using the purified protease, the blood stains were removed completely within 15 min in Set 1 [mixing of detergent (6 mg/mL) and purified enzyme]. On the other hand, it took 20 min when only the protease enzyme was used (Set 1). This result indicated that the addition of protease enzyme to a commercial detergent could significantly enhance the performance of the detergent in terms of removing the blood stains (Sathishkumar et al., 2015).

CHAPTER 9

ASCIDIAN CELLULOSE AS A POTENTIAL SOURCE OF INDUSTRIAL PRODUCTS

As extractive species, ascidians have been reported to make potential contribution in aquafeed formulation, biofuel production, IMTA, in dietary supplements, and so on. Apart from these, the cellulose of these ascidians is known to make significant contribution in the development of several industrial products. In this chapter, such ascidian–cellulose-based industrial products and the methods of their preparation are dealt with.

9.1 CHITOSAN

The waste shells of marine decapod crustaceans such as shrimps, crabs, and lobsters are the primary sources of chitin which yields the linear polysaccharide, namely, chitosan through enzymatic or chemical reactions. Chitosan is nontoxic, biodegradable, and biocompatible and is useful in many industrial applications including those of agriculture, food, biomedicine, cosmetics, and environmental protection. Certain species of ascidians including *P. nigra* have also been reported to be the sources of chitosan (Christomelba and Ananthan, 2016).

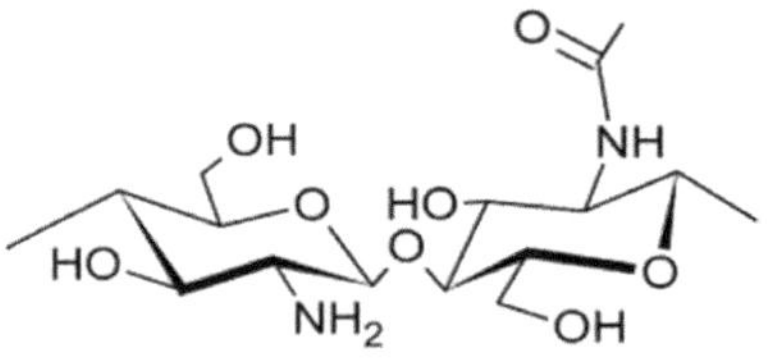

Chitosan

Biotechnological Potential of Ascidians: Wacky and Weird Sea Squirts. Ramasamy Santhanam (Author)

9.1.1 EXTRACTION OF CHITOSAN FROM THE TUNIC OF ASCIDIANS

The extraction of chitosan from the ascidians involves deproteinization, demineralization, and deacetylation. The air-dried ascidian tunic is thoroughly dried by keeping it in a hot air oven for 24 h at 60°C and this dried tunic is powdered using blender. Then the powdered tunic is treated with 5% sodium hydroxide solution and is kept at 60°C for 2 h in order to clear the proteins and other organic materials present in the tunic. After this treatment, the sample is washed again and again with distilled water until the solution becomes neutral and then it is allowed to dry to remove the excess water content present in the sample. This deproteinated sample is further treated with 1 M HCL solution for 24 h and is again washed with distilled water thoroughly until the solution comes to pH 7 and is dried to remove the water content from this de-mineralized sample which is then deacetylated by treated with 50% sodium hydroxide at 60°C for 8 h. At that stage also, the samples is repeatedly washed with distilled water until the solution gets pH 7. About 200 mL of 10% acetic acid is added to this sample and is kept for 12 h at room temperature. The dissolving chitosan is precipitated finally by adding 40% NaOH at pH 10 (Christomelba and Ananthan, 2016).

9.1.2 CHITOSAN DERIVED FROM PHALLUSIA NIGRA AND ITS ANTIBACTERIAL ACTIVITY AGAINST PATHOGENIC BACTERIA

The chitosan derived from the ascidian, *P. nigra* has been reported to possess antibacterial properties. At a concentration of 1 mg/mL, the chitosan inhibited the growth of several pathogenic bacteria with values of zone of inhibition ranging from 1 to 9.5 mm as given in Table 9.1.

TABLE 9.1 Antibacterial Activity of Chitosan Derived from *Phallusia nigra* on Pathogenic Bacteria

Pathogenic Species of Bacteria	Zone of Inhibition (mm)
Escherichia coli	1.5
Klebsiella oxytoca	9.5
Klebsiella pneumoniae	1.0
Vibrio cholerae	8.1
Vibrio parahaemolyticus	8.9
Vibrio vulnificus	4.8

Source: Adapted from Christomelba and Ananthan (2016).

9.1.3 OTHER COMMERCIAL APPLICATIONS

Chitosan has also been reported to possess several commercial applications. For example, it is amazing coagulating properties are of great use for treating waste waters. Further, the nontoxicity, biocompatibility, and antimicrobial activity of chitosan could lead to the production of chitosan-based films for food packaging which could be an alternative for the commercially available packaging materials which are known to hamper the environment. Chitosan films have also been considered as a potential local drug delivery system. Furthermore, the composite materials such as chitosan-hydroxyapatite and chitosan-alginate could be of great use for bone tissue engineering (Saha et al., 2014).

9.2 CHONDROITIN SULFATE

Chondroitin sulfate (CS), a complex, heterogeneous polysaccharide has been extracted from the tunic of the ascidian *S. clava*. This compound was found to show significant anti-inflammatory activity by inhibiting TNF-alpha-induced NF-kappaB activation and subsequent inducible nitric oxide synthase expressions and vascular cell adhesion molecule 1 by blocking the Akt signals (which promote tumor cell proliferation and metabolism) in human JB6 lymphoma cells. Further CS has also been reported to play an important role in in the treatment of osteoarthritis and preventing cardiovascular disease and colorectal cancer (Mazzucchelli et al., 2021).

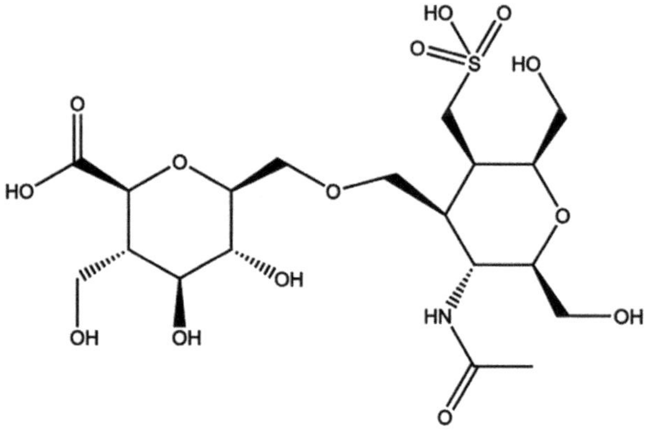

Chondroitin sulfate

9.3 ELECTROCATALYSTS

The edible ascidian, sea pineapple, *H. roretzi* has been reported to be a unique resource in the synthesis of electrocatalysts which are a kind of catalysts used to increase the rate of electrochemical reactions.

Principle: In the development of electrocatalysts, two industrial waste products are utilized to produce green hydrogen. A superior quality carbon could be created through carbonization of CNFs within the tunic of sea pineapples. This created carbon when mixed with pig's blood waste boasts electrical performances similar to the action of rare metals. Such catalysts may be utilized in advanced energy-related applications including polymer electrolyte fuel cells, metal-air batteries, and water electrolyzers.

Synthesis of carbon alloys: CNFs derived from the tunic of sea pineapple are extensively used nowadays as support for polymer materials, and their extremely crystalline structures possess ideal thermal and mechanical properties compared to the traditionally used CNFs derived from wood biomass. Before using, the CNFs are washed thrice using membrane-filtered water and it is collected through centrifugation (~3000 rpm). After purifying this water, a 2 wt% dispersion of CNFs is prepared in membrane-filtered water. The other component is the blood meal (BM) (a mixture of lipids and blood proteins) of pigs. Now the CNFs and BM are mixed in different weight ratios through hand grinding in water, and the mixtures so formed are dried at room temperature in vacuo. The dried samples are then pyrolyzed at 700, 800, or 900°C in vacuo for 3 h in an electric oven. A heating rate of 20°C/min is maintained. After this pyrolysis, the resultant sample is cooled to room temperature. The powder-like carbon alloys (CAs) are obtained after this pyrolysis process.

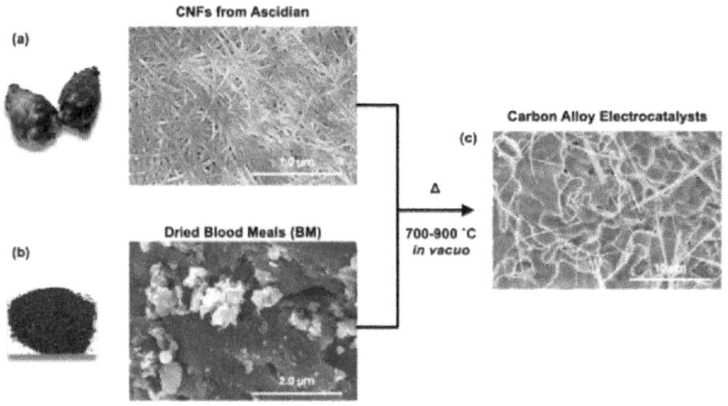

Use of sea pineapple shell CNFs and dried blood meal in the fabrication of catalysts: CNFs (a), BM (b), and the CAs (c).
Image credit: Reprinted with permission from Hiroshi Yabu et al. 2022.

9.3.1 JUSTIFICATIONS FOR THE USE OF ASCIDIANS AND BLOOD MEAL

The edible sea pineapple is an important food item in countries like Japan and Korea where they are produced at about 7000 t/year. This yield is believed to be enough in the production of CNFs at the rate of tens of tons/year. Further, compared to the wood biomass, the main advantage of CNFs from ascidian tunicates is that the CNFs may be harvested every year. Furthermore, the CNFs obtained from the ascidians, exhibit larger crystalline size, and greater crystallinity than wood-derived CNFs. The combination of ascidian-based CNFs and BMs is therefore considered to be a promising approach for preparing oxygen reduction reaction (ORR)/ oxygen evolution reaction (OER) electrocatalysts with high electron conductivity and high catalytic activity. Recent research findings have also reported that the pyrolysis of polydopamine-coated cellulose nanocrystals derived from the ascidians has demonstrated good OER and ORR. Above all, the development of such electrocatalysts from biomass resources would not only reduce CO_2 emissions but also utilize the shells of edible pineapples and livestock blood which may otherwise go as wastes.

It is also worth mentioning here that thousands of tons of dry blood wastes discarded annually from the livestock sector contaminate aquatic environments such as rivers, lakes, and streams. The discovery of electro-catalysts in this method will therefore help mitigating the pollution thereby a greener society with better sustainability could result.

9.4 CHEMICAL CELLULOSE

The ascidian tunic has been reported to be a potential source of chemical cellulose. In the development of chemical cellulose, the ascidian tunic undergoes prehydrolysis with acidic aqueous solution of H_2SO_4, digestion with alkali aqueous solution of $NaOH/Na_2S$, bleaching with aqueous NaOCl solution, and washing with acetone/water. The cellulose content the chemical cellulosed developed by this method was found to be 98 wt%. It is also worthy of mention here that Japanese Industrial Standard whiteness of the chemical cellulose is also about 98%. By the X-ray diffraction patterns and ^{13}C-NMR spectrum, it was found that the chemical cellulose prepared by the above method possessed cellulose Iβ crystal structure. Further, a new regenerated cellulose fiber was prepared from the above chemical cellulose by dry–wet spinning using *N*-methylmorpholine-*N*-oxide (NMMO)/water (87/13 wt%) as solvent. Compared to the commercially regenerated cellulose fibers, the new regenerated cellulose fiber prepared in this study has a higher ratio of wet-to-dry strength (<0.97) (Koo et al., 2002).

9.5 NANOFIBER-CARBON COMPOSITE FILMS

In order to preserve the operation of electronic gadgets, effective heat dissipation is mostly needed. For this purpose, thin and malleable thermal diffusion materials which can be used in thin-film devices could help a lot. In this regard, the cellulose of ascidians has been reported to play a significant role. The flexible nanocomposite films are produced with CNF matrix for which the sources are carbon fiber (CF) fillers and ascidians. These nanocomposite films are known to display superior anisotropic in-plane heat conduction. CNFs derived from the mantle of ascidians are believed to yield high thermal conductivity (about 2.5 W/mK) than the traditional polymers. Therefore the CNFs are considered very suitable for use as a heat-dissipating material. Further, apart from serving as superior thermal properties, the recyclability of these CF/CNF films is another important benefit. Furthermore, one could easily get the CFs by heating the ascidian-derived cellulose matrix and such CFs may also be reused. These outcomes can also be used as a framework for the engineering of 2D films with amazing heat dissipating patterns (https://www.tus.ac.jp/en/mediarelations/archive/20220922_6725.html)

9.6 PRESSURE-RESISTANT NANOFILTRATION MEMBRANE

A pressure-resistant nanofiltration membrane "Cellokon-AKH" could be manufactured from the cellulose of the tunic of sea peach *H. aurantium* using a simple, economical and environmentally friendly way. Scanning of the membrane surface with scanning electron microscopy (SEM), X-ray diffraction, and FTIR (Fourier transform infrared) spectroscopy, demonstrated that the membrane surface is an intertwined cellulose fiber network. While the lower surface was found to have both denser and looser areas, the upper surface displayed uniform morphology completely.

It is suggested that this CellokonAKH (dried form) membrane may be of great use in the filtration of substances with a molecular weight of more than 600 Da in various media (Kononova et al., 2022).

Membrane preparation: The biomaterial isolated from the selected ascidian is kept in formalin for a long-term storage. Before use, the tunic material is washed several times with water followed by ethanol. Then it is immersed in a medium of 96% ethanol. Before the preparation of the membrane, the cellulose is to be isolated from the biomaterial, namely, tunic. For this purpose, the biomaterial has to be pretreated through successive extractions with ethanol and acetone. The main purpose at this stage is to preserve the structural integrity of the mantle of the selected ascidian species, namely, *H. aurantium* by carefully removing the low-molecular weight substances. The sample so obtained is kept in 0.5 N. aqueous sodium hydroxide solution for 1 h and subsequently in 0.25 wt% NaOH solution at 100°C for 24 h. The resulting material is completely washed with water until the formation of neutral reaction, and subsequently with ethanol. Now a membrane results after the pressure-driven process of the liquid separation. This membrane is considered to be a porous anisotropic film of complex which is kept in various conditions, namely, either in the dry state or in 96% ethanol. Before storing, the membrane may also be dried in air under very normal conditions. The upper surface of the resulting membrane is grayish beige, and it is known to contain brown protrusions which are found evenly distributed over the entire surface of the membrane. However, the bottom of the film does not have the aforesaid color protrusions (Kononova et al., 2022).

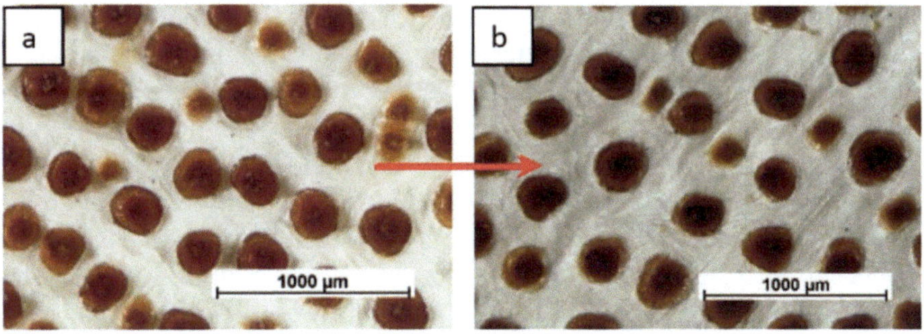

Optic microscope images of Cellokon-AKH film: (a–b) upper surface.

Source: Reprinted from Kononova et al., 2022. https://creativecommons.org/licenses/by/4.0/

9.7 FOOD TECHNOLOGY SECTORS

Food technology sectors such as food packaging, functional foods, and food stabilizers may also benefit from the cellulose and its derivatives, namely, cellulose nanocrystals (CNCs), and nanofibers of the ascidians. As proper food packaging is a necessity for protection, preservation of nutrition capacity, and food quality, the CNCs derived from the ascidians may be the best material for food packaging and their crystallinity and network structure offer an excellent oxygen barrier. Further, the presently used stabilizers are known to mitigate the instability of the processed foods and the ascidian-based cellulose could be the cheaper alternative to the stabilizers because they naturally emulsify and stabilizes. Further, ascidian-based cellulose composites such as hydroxyethyl may be used in milk drinks, ice cream, and frozen milk as a stabilizer to extend their shelf-life. Apart from processed foods, ascidian-based cellulose may also contribute its mite to the functional foods which are invariably dietary fiber foods that resolve digestive issues and prevent diseases. As ascidian cellulose is advantaged by its low calorie, it may be added to food to improve its fiber content and such foods are of great value to reduce obesity issues in humans. Though ascidian-based cellulose offers vast scope in the food technology sectors, considerable research is needed to put the ascidian-based cellulose into practice (Dunlop et al., 2020).

9.8 HUMIDITY SENSORS

Humidity sensors are of great use in food and drug storage/packaging, medical health, industrial manufacturing, agriculture, aviation, climatology, and so on. Ascidian-based cellulose and its derivatives are endowed with –OH functional groups and they offer good water retention and swelling performance. These features of the ascidian cellulose may be of great use in the manufacturing of green-humidity sensors. High-performance humidity sensors can therefore be fabricated from composites of cellulose with polymers/carbon materials. There are also reports that cellulose nanocrystals (CNCs) blended with modified poly(*N*-isopropyl acrylamide) copolymers to prepare a dual, humidity, and heat sensor (Zhu et al., 2022).

9.9 MISCELLANEOUS

Ascidian's larval bioadhesion properties offer vast scope in the development of man-made glues. Ascidians are also nowadays utilized as fishing bait and in the development of health supplement tablets. Further, several species of ascidians like *Halocynthia* spp. could also serve as ornamental animals in marine aquaria (Marques et al., 2022).

CHAPTER 10

ASCIDIANS AS THE POTENTIAL SOURCE OF SECOND-GENERATION BIOFUEL PRODUCTION

There is an urgent need to replace fossil fuels by biofuels and three generations of biofuels are presently known. Edible biomass, such as sugar, starch, animal fat, or vegetable oil are being used for the commercial scale production of first-generation biofuels; the feedstock for the production of second-generation biofuels consists of nonedible biomass, including municipal solid waste, waste vegetable oil, or lignocellulosic biomass which is mainly composed of agricultural wastes such as saw dust and agricultural residues like sugar cane bagasse, corncobs, and straw. Though this second-generation biofuel production is remaining at a pilot scale, several international commercial plants such as Finnish company UPM and Danish company Inbicon have started commercial production. While the Finnish company is known to generate 130 million liters of wood-based diesel annually, Inbicon is capable of producing 5.4 million liters of ethanol per year by processing 30 000 t of wheat straw. Further, anther Canadian company Iogen Corporation has generated 40 million liters of cellulosic ethanol from sugarcane bagasse from its commercial plant located at Brazil. In the second-generation biofuel production, search for additional feedstocks beyond forest biomass is also the need of the hour. In this regard, marine biomass such as seaweeds among marine plants and ascidians (marine invertebrate—chordates) offer vast scope in the bioethanol production. It is worth-mentioning here that these ascidians are the only marine animals possessing cellulose fibers linked with proteins in their outer covering, namely, tunic, an organic exoskeleton. This cellulose forms the starting material in the production of bioethanol. The

Biotechnological Potential of Ascidians: Wacky and Weird Sea Squirts. Ramasamy Santhanam (Author)

third-generation biofuel production which is from algae through photosynthesis has great potential for future applications but this technology largely depends on considerable improvement before it becomes economically viable (Hrůzová et al., 2020).

10.1 SECOND-GENERATION BIOFUEL PRODUCTION UTILIZING THE ASCIDIAN, *CIONA INTESTINALIS*

Among the commercially important species of ascidians, *C. intestinalis* has been reported to be the ideal species for the bioethanol production. In Rong, Norway, 100s of tons of this species are cultivated commercially and they are harvested in a size range of 15–20 cm (Apelgren et al. 2022). The different stages in the second-generation biofuel production utilizing this species are given below.

a. An individual b. Mariculture of *C. intestinalis*

Images credit: Reprinted with permission from a. Lars Kolby; b. Christofer & Eric

Production of tunic flakes from ascidian feed stock: The feedstock used in the production of second-generation biofuel production includes the species of ascidians produced by mariculture such as *Ciona intestinalis* (Hrůzová et al., 2020). The body of harvested *C. intestinalis* is first mechanically separated in two fractions: the tunic and the inner body part. During this process, the raw material is squeezed between two rollers, and the inner body was pressed out of the tunic. Both the inner body and the tunic are then separated by hand. Subsequently, the tunic is boiled at 100°C for 4 min without adding extra water. The material is then rinsed in fresh water for 4 min and pressed in a nylon cloth (approximately 0.3 mm openings) using a manual press. Then the filtered tunic is dried in a hot air oven at 80°C; and it is grinded into 3–75 mm tunic flakes which are then subjected to organosolv pretreatment, enzymatic saccharification, and ethanol fermentation.

Organosolv pretreatment: It is performed in an air-heated multidigester system with autoclave reactors. 110 g of dried and grinded whole tunic is mixed with 1.1 L of a 60% v/v ethanol solution and is used here. This treatment takes place at 175°C for 30 or 60 min. After this pretreatment, the reactor is cooled to less than 40°C. The pretreated solids are separated from the liquor by vacuum filtration, washed with 60% w/w ethanol, dried in the oven at 50°C overnight, and are finally stored in closed plastic bottles. The filtrate is processed in a rotary evaporator in order to recover the ethanol, while the aqueous solution is centrifuged to extract a solid precipitate which is collected, freeze-dried, and stored at room temperature. The remaining liquid is stored at 4°C. In laboratory studies, this organosolv pretreatment was found to enrich the tunic resulting in a cellulose content of 58.5% w/w, which is 3.4 times higher than the initial, untreated ascidian biomass.

Enzymatic saccharification: In this stage, the pretreated ascidian biomass is easily saccharified toward glucose and is then used as a carbon source for bioethanol production.

It is performed in 2-mL Eppendorf tubes containing 1 mL of 3% w/v dry solid in 50 mM citrate buffer (pH 5.0); and a thermomixer is used at 50°C for 48 h with 20 FPU/g solids of Cellic CTec2 commercial enzyme cocktail. Such samples are taken every 24 h, and the sugar profile is analyzed with high-performance liquid chromatography (HPLC). Samples that displayed the highest yields are then tested in ethanol fermentation trials.

Ethanol fermentation: The saccharified slurry is cooled at 35°C and is supplemented with $MgSO_4 \cdot 7H_2O$ (0.025 g/L), $(NH_4)_2HPO_4$ (0.5 g/L), and yeast extract (1 g/L). For the fermentation, *Saccharomyces cerevisiae* Ethanol Red is invariably used. For this fermentation process, yeast (lyophilized pellet) stored at 4°C is grown on YPD medium at 35°C for 24 h prior to inoculation. Then, the yeast cells are harvested by centrifugation (8000*g*, 10 min, 20°C) and are inoculated in the aforesaid saccharified slurry, at a final dry cell concentration of 1 g/L. The slurry is then incubated at 35°C, 200 rpm, for up to 240 h in an orbital shaker. Samples are taken every 24 h and are analyzed for the residual sugar and ethanol by HPLC. It is also reported that the fermentation yields were as high as 84.5%, with a final ethanol concentration of 34.9 g/L. Further, a maximal ethanol concentration of 38.7 g/L with a yield of 78.3% was achieved under the best conditions,

The levels of bioethanol production estimated with the different types of aquatic biomass are shown in Table 10.1.

TABLE 10.1 Bioethanol Production in the Different Types of Aquatic Biomass

Starter Material	Taxonomy	Yeast Strain Used	Ethanol Yield (g/L)
Ciona intestinalis	Ascidian	*Saccharomyces cerevisiae* Ethanol Red	38.7
Saccharina japonica	Green alga	*Saccharomyces cerevisiae* DK 410362	6.7
Laminaria japonica	Brown alga	*Debaryomyces occidentalis* (KCTC7196)	10.9
Gracilaria sp.	Red alga	*Saccharomyces cerevisiae* Wu-Y2	4.9
Eichhornia crassipes	Freshwater plant	*Saccharomyces cerevisiae* TY2	9.6

Source: Adapted from Hrůzová et al. (2020).

Table 10.1 data suggest that the ascidian biomass could be potential source for the second-generation bioethanol production. It is also suggested that the bioethanol production from ascidian biomass using organosolv pretreatment offers the vast opportunity to collaborate various processing industries thereby creating a circular economy. For example, mariculture of ascidians in shallow coastal waters improves local water quality and avoids marine pollution which is largely due to the filter-feeding habit of these ascidians and the removal toxic metals originated from industries and sewage waters. Second, removal of the inner parts of the ascidians during biofuel production helps for the production of finfish and shellfish aquafeeds.

CHAPTER 11

ASCIDIANS AS BIOREMEDIATING AGENTS OF MARINE POLLUTION

Bioremediation, a branch of biotechnology is the removal, reduction, or transformation of pollutants through biological processes. In aquatic environments, this phenomenon depends mostly microorganisms such as bacteria, fungi, and microalgae (microremediation) to degrade pollutants (https://en.wikipedia.org/wiki/Bioremediation). Most common instances of microbial bioremediation are removal of industrial spills, like oil spills. For oil spills, the processes, namely, bioaugmentation (i.e., introduction of microbes to the polluted sites) and biostimulation (i.e., stimulation of the existing microbial communities) for fast and efficient performance) are used. Further processes of microbial bioremediation include waste water effluent treatment in wastewater treatment plants, that is, removal of pollution-causing heavy metals like mercury, arsenic, and cadmium. Recent research findings have shown that the filter-feeding ascidians (sea squirts or tunicates play an important role in fighting against climate change as these organisms are capable of chemically sequestering carbon (i.e., absorbing unwanted carbon from the environment and storing it in their tissues). Further, these ascidians have been reported to concentrate harmful toxicants including heavy metals when they are present even at very low concentrations in the coastal waters (Tzafriri-Milo et al., 2019). Furthermore, several species of ascidians may serve as indicators of marine pollution. The ability of ascidians to accumulate certain heavy metals/trace elements from the seawater can be conveniently employed in order to define a suitable species of ascidian as indicator for some specific pollutants (Tzafriri-Milo et al., 2019; Papadopoulou and Kanias, 1977). Compared to other methods, bioremediation has been reported to be not

Biotechnological Potential of Ascidians: Wacky and Weird Sea Squirts. Ramasamy Santhanam (Author)

only more promising but are also less expensive way for the eradication of pollution in aquatic environments. The ascidians help in reducing marine pollution in several ways as detailed below.

Mitigating eutrophication: As filter feeders, the ascidians absorb their prey particles through pumping water through their oral siphon thereby they function as a "biological water filtration system." Through this process, these ascidians can remove harmful microalgae (e.g., harmful algal blooms, HABs) from eutrophic ecosystems (Kousoulaki et al., 2022).

Attenuating marine pollution: Ascidians are known to biodegrade pollutants; and they attenuate marine pollution by serving as bioremediators.

Bioindicators of water quality: Ascidians improve the water quality by removing harmful particles (pollutants/heavy metals) from the sea (https://thefishsite.com/articles/slimy-tunicates-a-new-aquaculture-adventure).

11.1 SPECIES OF ASCIDIANS SERVING AS BIOREMEDIATING AGENTS

11.1.1 CIONA INTESTINALIS

This species has been reported to reduce eutrophication by removing nitrogen and phosphate during its cultivation in shallow coastal waters (Hrůzová et al., 2020). Hassanzadeh (2011) reported that *C. intestinalis* may serve as marine pollution indicators (or bioremediators) due to its capability for accumulating certain trace elements like iron, zinc, and chromium. Based on the concentration factor of this species, it was found that this species can be used as indicator for iron (Papadopoulou and Kanias, 1977).

11.1.2 HALOCYNTHIA AURANTIUM

Several species of ascidians including *H. aurantium* have been reported to accumulate both essential trace elements necessary for their physiological functions and heavy metals with toxic characteristics. This method of bioaccumulating trace elements makes the ascidians as potential candidates for utilization as bioremediating agents (Gao et al., 2023).

11.1.3 *HERDMANIA PALLIDA, MICROCOSMUS EXASPERATUS, MICROCOSMUS SQUAMIGER, PHALLUSIA ARABICA,* AND *STYEKLA CANOPUS*

Radhalakshmi et al. (2014) reported that all these species concentrated heavy metals such as cadmium (Cd), copper, (Cu), lead (Pb), and vanadium (V), thereby they could serve as bioremediating agents. Among these species, *Phallusia arabica* displayed maximum biosorption of V with 21.23 ppm which is followed by Cu (4.57 ppm), Pb (1.54 ppm), and Cd (0.002 ppm). Least accumulation of Cd was however observed next to Pb. The Cd accumulation was from 0.0035 ppm (*H. pallida*) to 0.032 ppm in (*Styela canopus*). On the other hand, Pb was found accumulated in the range of 0.33 ppm (*H. pallida*) to 1.74 ppm (*S. canopus*) (Radhalakshmi et al., 2014).

Phallusia arabica

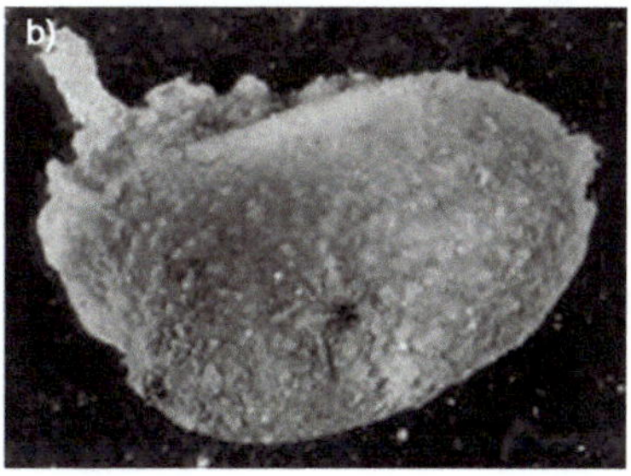

Styela canopus

Image credit: a. Julien Bidet for MDC Seamarc. https://creativecommons.org/licenses/by-sa/4.0/ b. Smithsonian Environmental Research Center - Styela_canopus_(I0909). https://creativecommons.org/licenses/by/2.0/

The values of concentrated heavy metals of these species against seawater levels and permissive levels of the different heavy metals prescribed by the World Quality Standard are shown in Tables 11.1–11.5.

TABLE 11.1 Concentration of Heavy Metals in Seawater and *Herdmania pallida* (in ppm)

Heavy Metals	Cadmium (Cd)	Copper (Cu)	Lead (Pb)	Vanadium (V)
Seawater	0.0028	0.0028	0.0026	0.0023
Herdmania pallida	0.0035	8.45	0.33	0.55
Bioaccumulation factor	1.25	4023	127	239
Permissive level (mg/kg)[a]	0.005 (FEPA, 2003)[b]	3 (FEPA, 2003)	2 (FEPA, 2003)	1.8 (CAC, 1993)[c]

Source: Adapted from Radhalakshmi et al. (2014). a-Prescribed by World Quality Standard; b-FEPA—Federation of European Producers of Abrasives; c-CAC—Codex Alimentarius Commission.

TABLE 11.2 Concentration of Heavy Metals in Seawater and *Microcosmus exasperatus* (in ppm)

Heavy Metals	Cadmium (Cd)	Copper (Cu)	Lead (Pb)	Vanadium (V)
Seawater	0.0028	0.0028	0.0026	0.0023
Microcosmus exasperates	0.0109	1.11	0.38	0.44
Bioaccumulation factor	3.9	529	146	191
Permissive level (mg/kg)[a]	0.005 (FEPA, 2003)[b]	3 (FEPA, 2003)	2 (FEPA, 2003)	1.8 (CAC, 1993)[c]

Source: Adapted from Radhalakshmi et al. (2014). a-Prescribed by World Quality Standard; b-FEPA—Federation of European Producers of Abrasives; c-CAC—Codex Alimentarius Commission.

TABLE 11.3 Concentration of Heavy Metals in Seawater and *Microcosmus squamiger* (in ppm)

Heavy Metals	Cadmium (Cd)	Copper (Cu)	Lead (Pb)	Vanadium (V)
Seawater	0.0028	0.0028	0.0026	0.0023
Microcosmus squamiger	0.0064	2.07	0.53	0.56
Bioaccumulation factor	2.28	985	204	243
Permissive level (mg/kg)[a]	0.005 (FEPA, 2003)[b]	3 (FEPA, 2003)	2 (FEPA, 2003)	1.8 (CAC, 1993)[c]

Source: Adapted from Radhalakshmi et al. (2014). a-Prescribed by World Quality Standard; b-FEPA—Federation of European Producers of Abrasives; c-CAC—Codex Alimentarius Commission.

TABLE 11.4 Concentration of Heavy Metals in Seawater and *P. hallusia arabica* (in ppm)

Heavy Metals	Cadmium (Cd)	Copper (Cu)	Lead (Pb)	Vanadium (V)
Seawater	0.0028	0.0028	0.0026	0.0023
Phallusia arabica	0.0199	4.57	1.54	22.23
Bioaccumulation factor	7.1	2176	592	9665
Permissive level (mg/kg)[a]	0.005 (FEPA, 2003)[b]	3 (FEPA, 2003)	2 (FEPA, 2003)	1.8 (CAC, 1993)[c]

Source: Adapted from Radhalakshmi et al. (2014). a-Prescribed by World Quality Standard; b-FEPA—Federation of European Producers of Abrasives; c-CAC—Codex Alimentarius Commission.

TABLE 11.5 Concentration of Heavy Metals in Seawater and *Styela canopus* (in ppm)

Heavy Metals	Cadmium (Cd)	Copper (Cu)	Lead (Pb)	Vanadium (V)
Seawater	0.0028	0.0028	0.0026	0.0023
Styela canopus	0.0322	3.38	1.74	3.55
Bioaccumulation factor	11.5	1609	669	1543
Permissive level (mg/kg)[a]	0.005 (FEPA, 2003)[b]	3 (FEPA, 2003)	2 (FEPA, 2003)	1.8 (CAC, 1993)[c]

Source: Adapted from Radhalakshmi et al. (2014). a-Prescribed by World Quality Standard; b-FEPA—Federation of European Producers of Abrasives; c-CAC—Codex Alimentarius Commission.

11.1.4 *MICROCOSMUS EXASPERATUS*

This species has been reported to absorb considerable amounts of toxic heavy metals thereby it could serve as a good biomonitor for some heavy metals. In the experimental study conducted using this species, it was found to concentrate maximum amount of manganese (209.1 mg/kg dw) and cadmium was the least accumulated one with the value of 0.46 mg/kg (Metri et al., 2019). The accumulated values of the different heavy metals are shown in Table 11.6.

TABLE 11.6 Concentration of Heavy Metals by *Microcosmus exasperatus*

Heavy Metal	Accumulated Level (mg/kg dw)	Certified Value and Organization
Manganese (Mn)	209.1	12 (FDA)
Zinc (Zn)	129.9	250 (ABIA)
Lead (Pb)	27.55	1.7 (FDA)
Chromium (Cr)	15.5	13 (FDA)
Copper (Cu)	14.4	150 (ABIA)
Nickel (Ni)	7.9	70 (FDA)
Vanadium (V)	6.3	4 (FDA)
Cadmium (Cd)	0.46	4 (FDA)

Notes: FDA, Food and Drug Administration; ABIA, Brazilian Association of Food Industries.

Source: Adapted from Metri et al. (2019).

Tzafriri-Milo et al. (2019) reported that these species accumulate several heavy metals in their body and tunic. Further, both the species have shown the capacity to concentrate large quantities of these heavy metals in their tissues (between 5000 and 2.5×10^6 times higher against the prevailing seawater concentrations).

11.1.5 *MICROCOSMUS EXASPERATES* AND *PHALLUSIA NIGRA*

Tzafriri-Milo et al. (2019) reported that these species accumulate heavy metals such as Al, V, Mn, Fe, Co, Ni, Cu, Zn, Cd, Ce, and Pb in the body and tunic. Further, both the species have shown an ability to accumulate large quantities of these heavy metals in their tissues (between ~5000 and 2.5×10^6 times higher compared to ambient seawater concentrations).

11.1.6 *MEAN VALUES (μg/g) OF HEAVY METAL CONCENTRATIONS IN THE TUNIC AND BODY OF* MICROCOSMUS EXASPERARUS (M.e.) *AND* PHALLUSIA NIGRA (P.n.)

Heavy Metal	*M.e.* Tunic	*M.e.* Body	*P.n.* Tunic	*P.n.* Body
Aluminum (Al)	4108	2763	525	5697
Cadmium (Cd)	0.1	0.06	0.2	0.1
Cerium (Ce)	9	4	1	8
Cobalt (Co)	7	2	2	3
Copper (Cu)	14	10	13	29
Iron (Fe)	4235	2564	716	5217
Manganese (Mn)	330	79	47	97
Nickel (Ni)	9	3	2	5
Lead (Pb)	11	4	5	6
Vanadium (V)	29	112	517	3426
Zinc (Zn)	25	60	12	39

Source: Adapted from Tzafriri-Milo et al. (2019).

11.1.7 *MICROCOSMUS VULGARIS* (=*Microcosmus sulcatus*)

According to the concentration factors, this species can be used as an indicator for the heavy metals such as selenium, chromium, zinc, and cobalt (Papadopoulou and Kanias, 1977; Hassanzadeh, 2011).

11.1.8 PHALLUSIA NIGRA

Based on the heavy metal concentrations, it is suggested that this species could serve as a valuable reference source for monitoring heavy metal contamination in Indian seas (Ali et al., 2015). Tzafriri-Milo et al. (2019) reported that this species concentrated the majority of heavy metals in its body.

11.1.9 STYELA PLICATA

This subtropical ascidian species has been reported to be very useful in the bioremediation of harmful bacteria and microalgal blooms appearing estuarine waters. The filtration rates of individuals of this species were experimented by exposing them to the microalgae *Nannochloropsis* sp. and bacteria *Escherichia coli* in concentrations of 10^5 and 10^6 cells/mL, respectively, for a period of 4 h. The results were as follows: An approximately −40 g sized individual ascidian exposed to 10^5 algal cells/mL was found to have an average filtration rate of 3.1 L/h and individual ascidian exposed to 10^6 algal cells/mL had an average filtration rate of 3.3 L/h, with 100% particle retention. On the other hand, the average filtration rates for the ascidians exposed to bacteria at 10^5 cells/mL were 4.7 L/h and were 2.3 L/h in the 10^6 cells/mL with 100% particle retention. Based on these results, it was estimated that 200 such ascidians could remove as much as 10^5 bacteria/mL each day from 22,600 L. It is also suggested that controlled densities of *S. plicata* may considerably reduce unwanted microalgae and bacteria in contaminated waters thereby improving water quality and mitigating eutrophication (Draughon, 2010). The heavy metal pollution study made in the coastal waters of eastern Aegean Sea, Aydn-Önen (2016) reported that *S. plicata* could be an ideal biological indicator for the biomonitoring of toxic heavy metals in the studied areas. They also found that the mean metal levels of heavy metals in *S. plicata* samples were in the order of Zn > Cu > Pb > V > Cd.

Further, very positive correlations have also been reported between Cu–Pb, Cu–Zn, and Pb–Zn in this ascidian samples. Furthermore, as *S. plicata* is also known to be a powerful accumulator of vanadium, the relatively low vanadium levels observed in the studied area may indicate the lack of anthropogenic sources of this metal in the sampled areas.

CHAPTER 12

MARICULTURE OF INDUSTRIALLY IMPORTANT ASCIDIANS

In the last few years, aquaculture, the culture of aquatic animals has undergone a significant worldwide growth. It is becoming the fast and growing food production industry in the world and it is estimated that about 50% of all fish consumed by humans is mainly from aquaculture practices (https://en.bdfish.org/2012/07/introduction-aquafeeds/).

Though several species of finfish and shellfish (i.e., shrimps, crabs, and mollusks) are being cultivated through coastal aquaculture and mariculture for human food and industrial requirements, several species are yet to be paid attention. Among such species, the cultivation of ascidians assume greater importance nowadays, owing to their importance as human food; and in the development of aqua and livestock feed, industrially important by-products, biofuel, and bio-remediators and therapeutic compounds with various bioactivities in general and anticancer in particular.

Advantages of ascidians in mariculture: Cultivation of ascidians poses a significant advantage because these animals grow faster and they can be grown underwater, thus not occupying large tracts of land which could otherwise be used to grow food or lush forests. Further, several species of ascidians are known to grow very quickly and they can be re-harvested two or three times a year (https://www.robaid.com/bionics/marine-animal-tunicate-can-be-used-both-as-biofuel-and-fish-food.htm).

12.1 INCORPORATION OF ASCIDIANS IN IMTA FRAMEWORKS

IMTA is a specialized type of aquatic polyculture, in which species occupying different trophic levels are integrated. Though considerable

Biotechnological Potential of Ascidians: Wacky and Weird Sea Squirts. Ramasamy Santhanam (Author)

interest is attributed to this IMTA, ascidians have not been adequately utilized in this framework. It is largely due to most publications which focused ascidians as pests due to their biofouling features and associated negative impacts on aquaculture facilities in general and on shellfish production in particular. Further, ascidian growth and fouling have caused considerable reduction in the size, growth and flesh weight of mussels. However, recent research findings reported on the positive impacts. For example, the biofouling ascidian, *C. intestinalis*, which has a fast-growing rate that contributes to organic sedimentation, is known to prefer unfouled sites, and it does not possess negative impacts on the associated bivalves or other organisms. Similarly, the presence of ascidian *S. clava* in an IMTA framework was found to optimize or enhance the growth of sea cucumber species, namely, *A. japonicus*. Experiments were conducted in the IMTA system using the ascidian *S. clava*, sea cucumber *A. japonicus* and microalgae and it was found that this IMTA framework reduced organic matter in the nearby sediments and mitigated the harmful bacterial populations and clean the water bodies from eutrophic nutrients like phosphorus and nitrogen Further, this framework demonstrated positive growth of sea cucumber with higher ascidian biomass (Marques et al., 2022).

12.2 HOW TO OVERCOME DIFFICULTIES IN INCORPORATING ASCIDIANS IN IMTA?

Use of ascidians as a component in IMTA may bring certain environmental impacts (e.g., biofouling) which may, however, be overcome by keeping specific materials such as long lines or PVC plates. These plates allow the settlement of ascidians without harming the aquaculture practice. Further, this arrangement may help fostering the production of biomass of ascidians and allowing for the easy removal of their biomass for multiple applications. Furthermore, the possibility of utilizing various ascidian species with shellfish, fish, or echinoderms (e.g., sea cucumbers) is also to be explored within IMTA frameworks (Marques et al., 2022).

Potential species of ascidians for mariculture: The species of ascidians cultivated in different countries for human food, livestock feed, medicinal compounds and industrial products include *Ascidiella aspersa*, *Botrylloides diegensis*, *B. schlosseri C. intestinalis*, *Corella willmeriana*, *E. turbinata*, *H. aurantium*, *H. roretzi*, and *S. clava* (Table 12.1).

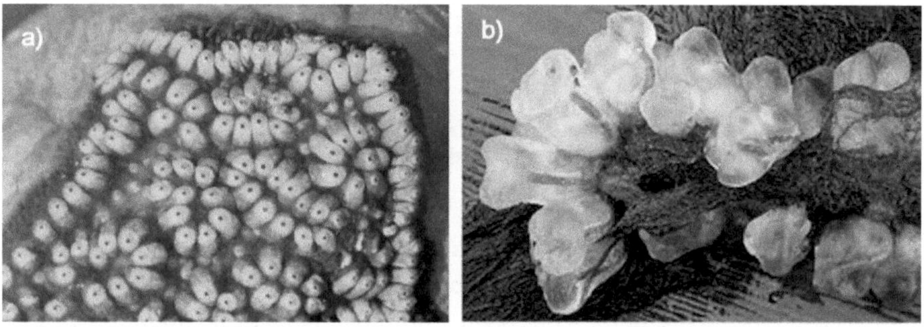

Botrylloides diegensis *Corella willmeriana*

Image credit: a) Connor. https://creativecommons.org/licenses/by/4.0/
b) Peter J. Bryant. https://nathistoc.bio.uci.edu/Tunicates/Colonial2.htm

TABLE 12.1 Commonly Cultured Species of Ascidians

Common Names	Species of Ascidian under Culture	Countries Where Cultured	References
–	*Botrylloides diegensis*	Switzerland	Wawrzyniak et al. (2021a)
Golden star tunicate	*Botryllus schlosseri*	Switzerland	Wawrzyniak et al. (2021a)
Vase tunicate	*Ciona intestinalis*	Bergen archipelago.	a
-do-	-do-	Worldwide	Joly et al. (2007)
Pacific transparent sea squirt	*Ciona savignyi*	Worldwide	Joly et al. (2007)
-do-	-do-	Canada, South Africa	b
Mangrove tunicate	*Ecteinascidia turbinata*		
Ice floc tuncate, sea peach, akaboya (Japan)	*Halocynthia aurantium*	Japan	Lambert et al. (2016)
Hoya or maboya (Japan), mung-geh (Korea), sea pineapple	*Halocynthia roretzi*	Japan, Korea	Lambert et al. (2016)
Clubbed tunicate, Asian tunicate, Mee-duh-duck (Korea), stalked sea squirt, rough sea squirt, or leathery sea squirt	*S. clava*	Korea	Lambert et al. (2016)
O-mahn-dee (Korea), pleated sea squirt	*Styela plicata*	Korea	Lambert et al. (2016)

Notes: [a]https://oceantunicell.com/news/an-update-from-our-aquaculture-facility/; [b]https://www.google.com/search?q=aquaculture+of++Ciona+intestinalis

12.3 METHODS OF CULTIVATING ASCIDIANS OF INDUSTRIAL IMPORTANCE

Industrially important ascidians are cultivated by two main methods, namely, culture in flow-through systems and culture in closed systems with periodic water changes.

Culture in a flow-through system: In this method, filtered natural seawater is used from the sampling site which yields almost natural conditions including the feeding diet. This method has been reported to isolate and maintain ascidian strains for long-term culture in confined and safe conditions. For successful maintenance, this method needs either a close proximity to seawater source.

Culture in a closed system with periodic water changes: Aquaculture in a closed system with routine water changes: This method needs seawater from the site of sampling for short-term aquaculture of organisms. Further, this method may also involve artificial seawater (ASW).

12.4 IN-LAND IN-LAB MARINE RECIRCULATING ARTIFICIAL SEAWATER SYSTEM FOR LONG-TERM CULTURE AND BREEDING OF COLONIAL *BOTRYLLOIDES DIEGENSIS* AND *BOTRYLLUS SCHLOSSERI*

The colonies of *B. diegensis* are composed of several genetically identical adults, called as zooids which are firmly attached in a gelatinous matrix, called tunic. Individual zooids are of about 23 mm in length. The life cycle of this species commences seasonally from a tadpole larval stage. After a free-swimming phase, this tadpole larva gets settled and metamorphosed into a sessile adult animal called as oozooid. In this culture method, the recirculating system keeps running on ASW and employs macroalgae pastes as diets for the growing ascidians. The aquarium designed for the culture of the ascidians allows four various culture conditions to suit the breeding of the required ascidian species (Wawrzyniak et al., 2021a).

12.5 RECIRCULATING AQUACULTURE SYSTEM FOR THE IN-LAND AQUACULTURE OF COLONIAL ASCIDIANS

Recirculating aquaculture system: The main advantage of this recirculating system is the reduction of periodic water changes to minimal amounts.

Further, in these closed systems, the wastes such as NH_4, NO_2, and NO_3 produced by the growing ascidians accumulate over time and they need to be continuously purified. This aquaculture system is designed at three levels around a stainless stand in order to optimize or maximize the use of gravity to facilitate the easy circulation of the artificial seawater in the culture system. The food culturing tanks situated on the top of the culture system distribute their content into the ascidian tanks present in the center. The water filtration unit kept at the bottom of the apparatus has both biological and mechanical filters to attain a minimum of weekly water change (5%) of the total amount of 450 L of ASW originally present in the culture system. Filtration is facilitated by a particle filter, a denitrifyer filter, an algae scrubber, a protein skimmer, a moving bed filter (or upflow filter), live stones, and a UV-C unit (Wawrzyniak et al., 2021a).

Operation of recirculating aquaculture system: Four 40-L specimen tanks are chosen with one compartment in each tank. Every tank has a heater which is located at the bottom of the tank. Mixing of the water and aeration are facilitated by a circulation pump which is located on the tank's back side. These tanks are fully closed in order to make a light-free environment and are with a front opening for cleaning. Water enters into the tank through a silicone tube, and it goes out of the tank through an outlet present in the tank's back side. A fleece is kept at the bottom of the tanks to collect all solid wastes. Each tank is provided with two lights and the photoperiod and intensity of each LED unit is controlled by a central control unit. Ascidians are kept in the tanks on hangers made of plastic hooks, plastic tubes, and badge grippers. The central filtration unit (sump) is located on the ground. Through a fleece filter, water is drained from the tanks into an aerated moving-bed biofilter and a ceramic-beads biofilter; and finally into a main sedimentation tank. The latter is attached to an outer protein skimmer (with submerged pump), a nitrate filter, an algal scrubber and a refugium which is provided with macroalgae and live cured stones. The sedimentation tank is united to the chief circulation pump, which provides water from the culture system to the cooling unit, refugium in a closed loop, algal scrubber and also the ascidian tanks through a UV sterilizer. A total of four 10-L cylindrically designed feeding tanks are kept above the specimen tanks. Each of these tanks is made of PVC pipes with 20 cm diameter and 30 cm length. The content of the tanks gets stirred through the injection of air from a microporous tube kept at the bottom via the center of the cover of the tank. Food is sent out from each tank by gravity. Air is sent directly through 4 mm silicone hoses from the main air pump via a custom-built manifold. Oxidation reduction potential, temperature, pH, and salinity, are all measured often in the main sedimentation tank. Water level is monitored with two optical sensors the sedimentation tank. Other parameters of water quality such as pH, NH_4, NO_2, NO_3, PO_4 are monitored manually (Wawrzyniak et al., 2021a).

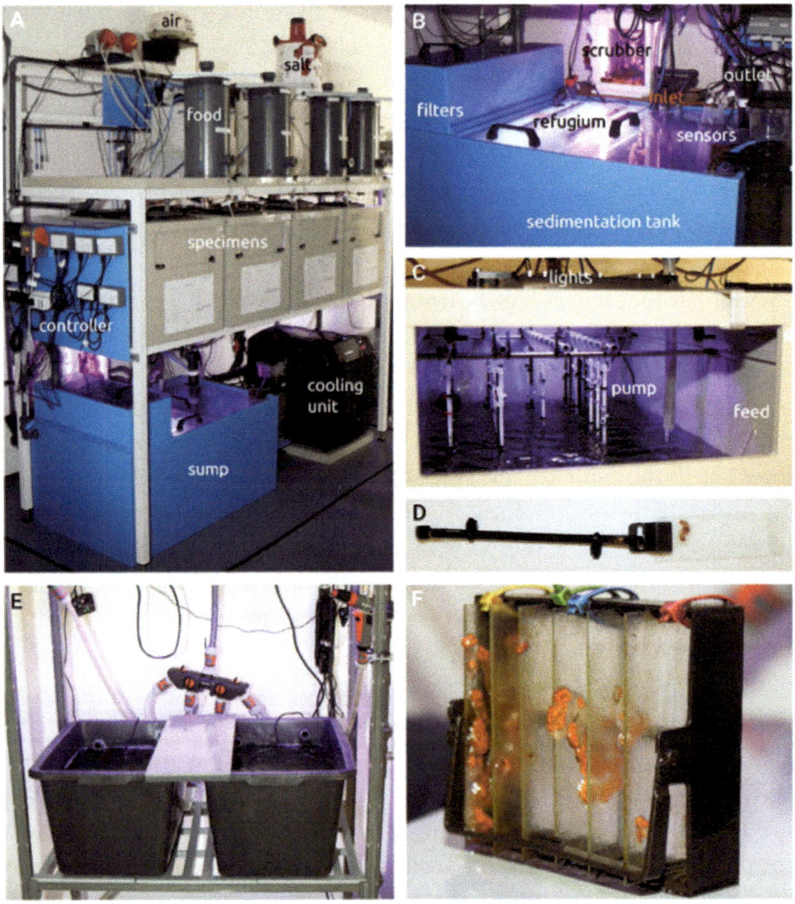

Aquaculture setup of the colonial ascidians. Image of the prototype development (A), inside the sump view (B), vertical culture system of colonial ascidians (C), colony hanger (D), quarantine containers (E), and staining racks for quarantine (F). Artificial seawater based long-term culture of colonial ascidians (Marta K. Wawrzyniak, Lluìs Albert Matas Serrato, Simon Blanchoud. *Developmental Biology*, **2021,** *480*, 91–104).

Images credit: Reprinted with permission from Wawrzyniak, Serrato, Blanchoud.

Breeding colonial tunicates: Collected ascidians are first transferred to a 1-L wide-mouthed cylindrical container. Animals are then transported to the laboratory. Upon arrival, wild specimens are introduced to 2-L tanks which are filled with ASW with more or less same salinity and pH of the collected seawater. Animals are placed in quarantine tanks and are provided with feed twice daily. This feed is mixed with 1 mL of feeding mix and are left 24 h to condition. On the next day, colonies are subcloned to the microscopy glass slides (numbered) with a single-edged razor blade. These slides are then placed in staining racks made of plastic at the rate of seven slides per rack which is allowed to stand at the base of the quarantine tank. Colonies are then left for 48 h to settle. For about 30 L setup, this will take about a week for two dozen colonies, The colonies that detached are again subcloned into the first quarantine and this process is repeated until all colonies are either attached or decayed. Colonies are then acclimatized to the water-quality parameters of the created recirculating system through gradually adjusting the parameters of the quarantine for about two weeks. The breeding water-quality parameters are chosen to be more or less close to the in situ environmental parameters of the colonies. Colonies are then allowed to spend another two weeks in a second quarantine before they are transferred to the main system where they are known to proliferate actively. The attached colonies are then placed in the second quarantine tank. Colonies that are detached are subcloned again into the first quarantine and the process is repeated until all colonies are either attached or decayed. Colonies are then acclimatized to the environmental parameters of the created recirculating system by gradually adapting the parameters of the quarantine over two weeks. The breeding environmental parameters are chosen to be more or less similar to the in-situ water-quality conditions of the colonies. Colonies are then allowed to spend another two weeks in a second quarantine before they are transferred to the main system where they are known to proliferate actively. It was found that about 50% of the collected ascidians could be successfully transferred to the created recirculating system (Wawrzyniak et al., 2021a).

12.6 MARICULTURE OF *HALOCYNTHIA RORETZI* AND *STYELA CLAVA*

The mariculture of *Halocynthia* spp. and *Styel*a spp. was started in Korea and Japan in order to meet the product demand. Korea started the farmed production of *Halocynthia roretzi* in 1982 and a total of 16,966 t (6,994 t of

H. roretzi and 9,972 t of *S. clava*) and subsequently Japan stared the farming of *H. roretzi* (Lambert et al., 2016). In both Korea and Japan, these species are commonly cultivated on long line systems. *S. clava* seed are collected from the shallow subtidal areas and are then transplanted for grow-out on longlines where the seed grow to marketable size in about a year. Individuals of *H. roretzi* are spawned in hatcheries; and the larvae attach to ropes which are then kept in the sea hanging from long horizontal lines. In this species, the attached larvae attain harvestable size of 22 cm in height in two years in Korea and in about three years in Japan. Though a water temperature of 2–26°C is maintained for the cultivation of *H. roretzi*, the optimum growth was reported in the temperature range of 8–13°C (Lambert et al., 2016). Changes in environmental factors such as water temperature, salinity, or dietary plankton have been reported to cause the disease soft tunic syndrome in cultured *H. roretzi* and this caused mass mortality and associate decreased production and huge economic loss in Korea. Symptoms of this disease are thinner and separated siphonal spines, thinning of the tunic cuticle, and lowered density of tunic fibers which eventually result in rupturing of the tunic and death. The kinetoplastid protist parasite, namely, *Azumiobodo hoyamushi* has been reported to be he causative agent for soft tunic syndrome (Lambert et al., 2016).

12.7 MARICULTURE OF THE MANGROVE ASCIDIAN, *ECTEINASCIDIA TURBINATA*

E. turbinata is an Atlanto-Mediterranean species growing commonly in mangroves and coastal meadows (Naranjo et al., 2001). Since most of its secondary metabolites occur at low concentrations, large amounts of biomass of this species are needed for the development of new drugs. As large-scale harvesting of this species would expensive and environmentally destructive, alternative supply method, that is, mass culturing of this species is a must. In aquaculture, this species has shown not only enhanced growth rates but also increased biosynthesis of required bioactive compounds which are known to possess antitumor properties.

12.8 *IN-VITRO* CULTURE OF *ECTEINASCIDIA TURBINATA*

One preferred method of supplying sustainable and sufficient quantities of bioactive metabolite of this species is its *in-vitro* culture, in which this species is kept in tanks on land and is grown with controlled conditions. However, it is very important to create suitable feeding regimes for the successful in-vitro culture. Only such conditions could enhance the growth and biosynthesis production of the required metabolites.

12.9 EXPERIMENTAL SETUP

For each experiment, approximately 15 large colonies of 100–200 zooids each were collected from the natural environment at a depth of 10–50 cm. Each of such collected colonies had both stolon and zooids and was also free of other macro-fouling organisms. These collected colonies were then separated into smaller colonies of about 20 zooids with common stolon using a scalpel. Then, with the help of a thin nylon line, these small colonies were individually tied to 36-cm nylon mesh pieces in which both the mesh diameter and hole size were of 1 mm. Now, these small colonies are tied with the stolon which rested against the nylon mesh to enhance attachment. Preliminary experiments conducted showed that this species of ascidian attached quickly and grew well on nylon mesh. An extra length of nylon line was also tied to the top of each mesh piece to make a loop which was threaded over a plastic rod laid over each aquarium so that each of such colonies could hang comfortably in midwater. In order to keep the colonies hanging vertically, a small weight was also secured to the bottom of each mesh piece. In each aquarium, three of the above-prepared colonies were kept at 5 cm intervals on the plastic rod.

12.10 EXPERIMENTAL AQUARIA

The aquaria used for the experiment were of 252,941 cm in size with 15 l of natural seawater. In order to remove unwanted particles, this seawater was first passed through a series of UV lights and filters. In each aquarium, the seawater was replaced with clean seawater 5–6 times each week. This procedure helped largely to remove any uneaten food cells and feces, besides prevented preventing high ammonia levels. Dissolved ammonia was kept at 0.01 mg/L throughout each experiment. Other environmental factors such as water temperature, salinity, and pH were kept almost constant throughout the experiment at 25.8°C, 33×, and 8.1, respectively. This procedure is to keep the natural conditions. The seawater was also aerated constantly with the help of two air-stones kept in opposite corners of each aquarium. The air-stones were however kept away from all experimental ascidians so as to prevent their direct exposure to the air bubbles. In each aquarium, the seawater was aerated at the rate of 220 mL/s in order to provide sufficient oxygen for respiration and to facilitate good water circulation so as to keep the added food particles always in suspension (Duckworth et al., 2004).

12.11 FEEDING THE ASCIDIAN COLONIES

The colonies are fed with three species of microalgae, namely, *Nannochloropsis* sp., *Chaetoceros gracilis*, and *Isochrysis galbana* either singly or in combination at a density of about 80,000 cells/mL which is almost a natural concentration. The growth studies made showed that maximum growth was noticed in colonies fed with monospecic diet of *Isochrysis galbana* and a multispecic diet of *Chaetoceros gracilis* and *Isochrysis galbana* and it is suggested that these two diets would adequately satisfy the nutritional needs of this ascidian. In order to assess the optimal diet concentration for this in-vitro culture, colonies were first fed with both the diets at 1, 2, and 4 fed the natural-level diets showed little grown while colonies fed with higher concentrations doubled in their size. Further, this in-vitro culture was also found to promote target metabolite biosynthesis as farmed colonies had four times the normal concentration of the metabolite, namely, ecteinascidins (Duckworth, 2004).

REFERENCES

Aiello, A.; Fattorusso, E.; Menna, M.; Iuvone, T. Sulcatin. A novel antiproliferative *N*-methylpyridinium alkaloid from the ascidian *Microcosmus vulgaris. J. Nat. Prod.* **2000,** *63,* 517–519.

Ali, H.A.J.; Tamilselvi, M.; Akram, A.S.; Arshan, M.L.K.; Sivakumar, V. Comparative study on bioremediation of heavy metals by solitary ascidian, *Phallusia nigra,* between Thoothukudi and Vizhinjam ports of India. *Ecotoxicol. Environ. Saf.* **2015,** *121,* 93–99.

Ananthan, G.; Karthikeyan, M.M.; Prabhu, S.A.; Raghunathan, C. Studies on the seasonal variations in the proximate composition of ascidians from the Palk Bay, Southeast coast of India. *Asian Pac. J. Trop. Biomed.* **2012,** *2,* 793–797.

Apelgren, P.; Sämfors, S.; Säljö, K.; Mölne, J.; Gatenholm, P.; Troedsson, C.; Thompson, E.M.; Kölby, L. Biomaterial and biocompatibility evaluation of tunicate nanocellulose for tissue engineering. *Biomater. Adv.* **2022,** *137,* 212828.

Aydın-Önen, S. Styela plicata: A new promising bioindicator of heavy metal pollution for eastern Aegean Sea coastal waters. *Environ. Sci. Pollut.Res.Int.* **2016,** *23,* 21536–21553.

Behrendt, L.; Larkum, A.W.D.; Trampe, E.; Norman, A.; Sørensen, S.J.; Ku¨hl, M. Microbial diversity of biofilm communities in microniches associated with the didemnid ascidian *Lissoclinum patella. ISME J.* **2011,** 1–16.

Berrill, N.J. Studies in tunicate development Part IV—Asexual reproduction. *Philos. Trans. R. Soc. Lond., B, Biol. Sci.* **1935,** *225,* 327–379

Bugaj, M. Species Distribution and Abundance of Bering Sea Tunicates with Implications for Coastal Food Security. Master's Thesis, WWU Graduate School Collection, **2022.**

Castilla, J.C.; Defeo, O. Latin American benthic shellfisheries: Emphasis on co-management and experimental practices. *Rev. Fish Biol. Fish.* **2001,** *11,* 1–30.

Chan, S.T.S.; Pearce, A.N.; Page, M.J.; Kaiser, M.; Copp, B.R. Antimalarial β-carbolines from the New Zealand ascidian *Pseudodistoma opacum. J. Nat. Prod.* **2011,** *74,* 1972–1979.

Christomelba, D.C.; Ananthan, G. Chitosan derived from the tunic of ascidian *Phallusia nigra* (Savigny, 1816) showing antibacterial activities and its characterization. *Int. J. Med. Sci.* **2016,** *8,* 91–95.

Cihangir, H.A.; Muñoz, A.I.; Papadopoulou, M.A.P.; A. Esplá, A.A.R.; Yılmaz, E.C. *Microcosmus polymorphus* Heller, 1877 (Tunicata: Ascidiacea: Pyuridae)—A new addition to the fauna of the Turkish coasts. *Turkish J. Fish. Aquat. Sci.* **2011,** 11, 161–163.

Cruz-Monserrate, Z.; Vervoort, H.C.; Bai, R.; Newman, D.J.; Howell, S.B.; Los, G; Mullaney, J.T.; Williams, M.D.; Pettit, G.R.; Fenical, W.; Hamel, E. Diazonamide A and a synthetic structural analog: Disruptive effects on mitosis and cellular microtubules and analysis of their interactions with tubulin. *Mol. Pharmacol.* **2003,** *63,* 1273–1280.

de Menezes, C.B.A.; Afonso, R.S.; de Souza, W.R.; Parma, M.; de Melo, I.S.; Zucchi, T.D.; Fantinatti-Garboggini, F. *Gordonia didemni* sp. nov. an actinomycete isolated from the marine ascidium *Didemnum* sp. *Antonie van Leeuwenhoek* **2016,** *109,* 297–303.

Dou, X.; Li, X., Yu, H.; Dong, B. Dual roles of ascidian chon-dromodulin-1: Promoting cell proliferation whilst suppressing the growth of tumor cells. *Mar. Drugs* **2018**, *16*, 59.

Dou, X.; Dong, B. Origins and bioactivities of natural compounds derived from marine ascidians and their symbionts. *Mar. Drugs* **2019,** *17*, 670.

Draughon, L.D. The Ascidian, *Styela plicata* as a Potential Bioremediatory of Bacterial and Algal Contamination of Marine Eestuarine Waters. Ph.D.Thesis, Florida Atlantic University, **2010.**

Duckworth, A. R.; Samples, G.A.; Wright, A.E.; Pomponi, S.A. In vitro culture of the ascidian *Ecteinascidia turbinata* to supply the antitumor compounds ecteinascidins, *Aquaculture* **2004**, *241*, 427–439.

Dunlop, M.J.; Clemons, C.; Reiner, R.; Sabo, R.; Agarwal, U.P.; Bissessur, R.; Soioudiasli, H.; Carreau, P.; Acharya, B. Towards the scalable isolation of cellulose nanocrystals from tunicates. *Sci. Rep.* **2020**, *10*, 19090. https://doi.org/10.1038/s41598-020-76144-9

Edler, M.C.; Fernandez, A.M.; Lassota, P.; Ireland, C.M.; Barrows, L.R. Inhibition of tubulin polymerization by vitilevuamide, a bicyclic marine peptide, at a site distinct from colchicine, the vinca alkaloids, and dolastatin 10. *Biochem. Pharmacol.* **2002,** *63,* 707–715.

Fulgione, D.; Rippa, D.; de Luca, A.; Milone, M. Molecular characterization of 18S rDNA partial sequence in *Microcosmus* (Stolidobranchiata, Pyuridae). *Mediterr. Mar. Sci.* **2006**, *7,* 5–10.

Gandra, M.; Cavalcante, M.; Pavao, M. Anticoagulant sulfated glycosaminoglycans in the tissues of the primitive chordate *Styela plicata* (Tunicata). *Glycobiology* **2000,** *10,* 1333–1340.

Gao, P.; Khong, H.Y.; Mao, W.; Chen, X.; Bao, L.; Wen, X.; Xu, Y. Tunicates as sources of high-quality nutrients and bioactive compounds for food/feed and pharmaceutical applications: A review. *Foods* **2023,** *12,* 3684.

Gasparini, F.; Ballarin, L. Comparative reproduction. *Encyclopedia of Reproduction* (second edition), Elsevier, **2018.**

Hassanzadeh, M. Composition and Application Potentials of Scandinavian Tunicates. Master Thesis Report, KTH Chemical Science and Engineering, **2011.**

Holland, L.Z. Tunicates. *Curr. Biol.* **2016,** *26,* R146–R152.

Hong, S.; Kwone, J.; Leea, J.; Leea, S.; Leea, A.Y.; Choa, W.; Bat-Erdenef, M.; Choif, B.; Cho, M. Ascidian tunicate extracts attenuate rheumatoid arthritis in a collagen-induced murine model. *Nat. Prod. Commun.* **2014,** *9,* 847–851.

Hrůzová, K.; Matsakas, L.; Karnaouri, A.; Norén, F.; Rova, U.; Christakopoulos, P. Second-generation biofuel production from the marine filter feeder *Ciona intestinalis. ACS Sustainable Chem. Eng.* **2020,** *8,* 8373–8380.

Joly, J.; Kano, S.; Matsuoka, T.; Auger, H.; Hirayama, K.; Satoh, N.; Awazu, S.; Legendre L.; Sasakura, Y. Culture of *Ciona intestinalis* in closed systems, *Dev. Dyn.* **2007,** *236,* 1832–18440.

Karamanou, K.; Espinosa, D.C.R.; Fortuna-Costa, A.; Pavão, M.S.G. Biological function of unique sulfated glycosaminoglycans in primitive chordates. *Glycoconj J.* **2017**, *34,* 277–283.

Karthikeyan, M. M.; Anathan, G.; Ali, A. J. Nutritional values of solitary ascidian *Microcosmus exasperatus* Heller, 1878 (Ascidiacea: Pyuridae), *Glob. Vet.* **2010,** *4,* 255–259.

Kijajic, Z.; Dogovic, N.; Gasic, M.J. Sterols in adriatic sea ascidians. *Comp. Biochem. Physiol.* **1983,** *75B,* 519–521.

Kononova, S.V.; Khripunov, A.K.; Romanov, V.N.; Orekhov, A.S.; Mikhutkin, A.A.; Vlasova, E.N.; Lukasov, M.S.; Klechkovskaya, V.V. Animal cellulose with hierarchical structure isolated from *Halocynthia aurantium* tunic as the basis for high-performance pressure-resistant nanofiltration membrane. *Membranes.* **2022,** *12,* 975.

Koo, Y.; Young-Soo, W.; You, S.; Kim, H. Preparation and properties of chemical cellulose from ascidian tunic and their regenerated cellulose fibers. *J. Appl. Polym. Sci.* **2002**, *85,* 1634–1643.

Kousoulaki, K.; Sveen, L.; Norén, F.; Espmark, A. Atlantic salmon (Salmo salar) performance fed low trophic ingredients in a fish meal and fish oil free diet. *Front. Physiol.,* **2022,** https://doi.org/10.3389/fphys.2022.884740

Kozlowski, E.O.; Lima, P.C.; Vicente, C.P.; Lotufo, T.; Bao, X.; Sugahara, K.; Pavão, M.S.G. Dermatan sulfate in tunicate phylogeny: Order-specific sulfation pattern and the effect of [→4IdoA(2-Sulfate)β-1→3GalNAc(4-Sulfate)β-1→] motifs in dermatan sulfate on heparin cofactor II activity. *BMC Biochem.* **2011,** *12,* 29.

Kumaran, N.S.; Bragadeeswaran, S. Nutritional composition of the colonial ascidian *Eudistoma viride* and *Didemnum psammathodes. Biosci. Biotechnol. Res. Asia* **2014,** *11,* 331–338.

Kurn, U.; Rendulic, S.; Tiozzo, S.; Lauzon, R.J. Asexual propagation and regeneration in colonial ascidians. *Biol. Bull.* **2011,** *221,* 43–61.

Lambert, G.; Karney, R.C.; Rhee, W.Y.; Carman, M.R. Wild and cultured edible tunicates: a review. *Manag. Biol. Invasions.* **2016**, *7,* 59–66.

Le, V.H.; Inai, M.; Williams, R.M.; Kan, T. Ecteinascidins. A review of the chemistry, biology and clinical utility of potent tetrahydroisoquinoline antitumor antibiotics. *Nat. Prod. Rep.* **2015,** *32,* 328–347.

Lee, K.H.; Hong, B.I.; Jung, B.C.; Cho, H.S.; Park, C.S.; Jea, Y.G. Seasonal variations of nutrients in warty sea squirt (*Styela clava*). *J Korean Soc. Food Sci. Nutr.* **1995**, *2*, 268–273.

Li, J.L.; Xiao, B; Park, M.; Yoo, E.S.; Shin, S.; Hong, J.; Chung, H.Y.; Kim, H.S.; Jung, J.H. PPAR-γ agonistic metabolites from the ascidian *Herdmania momus. J. Nat. Prod.* **2012b,** *75,* 2082–2087.

Li, C.; Li, D.; Yin, X.; Chang, Y. Analysis and evaluation of the nutritional composition of ascidian *Haloeynthia roretzi. J. Dalian Ocean Univ.* **2007**, *22,* 347–351.

Makar'eva, T.N.; Ponomarenko, L.P.; Stonik, V.A. Steol composition of an ascidian *Polycitor* sp. *Chem. Nat. Compd.* **1995,** 31, 637–638.

Manríquez, P.H.; Castilla, J.C. Roles of larval behaviour and microhabitat traits in determining spatial aggregations in the ascidian *Pyura chilensis. Mar. Ecol. Prog. Ser.* **2007,** *332,* 155–165.

Marques, L.; Calado, R.; Lillebo, I. Potential of ascidians as extractive species and their added value in marine integrated multitrophic aquaculture systems—From pests to valuable blue bioresources. *Front. Mar. Sci.* **2022**, *9,* https://doi.org/10.3389/fmars.2022.849870

Marques, L.; Domingues, M.R.; da Costa, E.; Abreu, M.H.; Lillebø, A.I.; Calado, R. Screening for health-promoting fatty acids in ascidians and seaweeds grown under the influence of fish farming activities. *Mar. Drugs* **2021,** *19,* 469.

Matos, A.; Antunes, A. Symbiotic associations in ascidians: Relevance for functional innovation and bioactive potential. *Mar. Drugs* **2021,** *19,* 370.

Mazzucchelli, R.; Rodríguez-Martín, S.; García-Vadillo, A.; Gil, M.; Rodríguez-Miguel, A.; Barreira-Hernández, D.; García-Lledó, A.; de Abajo, F.J. Risk of acute myocardial infarction among new users of chondroitin sulfate: A nested case-control study. *PLoS One* **2021,** *16*, e0253932.

McDonald, L.A.; Capson, T.L.; Krishnamurthy, G.; Ding, W.-D.; Ellestad, G.A.; Bernan, V.S.; Maiese, W.M.; Lassota, P.; Kramer, R.A.; Ireland, C.M. Namenamicin, a new enediyne antitumor antibiotic from the marine ascidian *Polysyncraton lithostrotum*. *Am. Chem. Soc.* **1996,** *118*, 10898–10899.

Meenakshi, V.K.; Gomathy, S.; Senthamarai, S.; Paripooranaselvi, M.; Chamundeswari, K.P. Analysis of vitamins by HPLC and phenolic compounds, flavonoids by HPTLC in *Microcosmus exasperatus*. *Eur. J. Zool. Res.* **2012,** *1,* 105–110.

Metri, R.; Soares, G.C.B.; Guilherme, P.D.B.; Roveda, L.F. The ascidian *Microcosmus exasperatus* as bioindicator for the evaluation of water quality in estuaries. *Int. J. Adv. Res.* **2019,** *7,* 174–185.

Michibata, H. The mechanism of accumulation of vanadium by ascidians: Some progress towards an understanding of this unusual phenomenon. *Zool. Sci.* **1996**, *13,* 489–502.

Michibata, H.; Terada, T.; Anada, N.; Yamakawa, K.; Numakunai, T. The accumulation and distribution of vanadium, iron and manganese in some solitary ascidians. *Biol. Bull. In.* **1986,** *171*, 672–681.

Mizuta, S.; Isobe, S.; Yoshinaka, R. Existence of two molecular species of collagen in the muscle layer of the ascidian (*Halocynthia roretzi*). *Food Chem.* **2002**, *79*, 9–13.

Nakauchi, M. Asexual Development of ascidians: Its biological significance, diversity, and morphogenesis. *Am. Zool.* **1982,** *22,* 753–763.

Nam-Do, C.; Jiting, Z.; Byung-Dae, C.; Hong-Soo., R. Shelf life of bottled sea squirt *Halocynthia roretzi* meat packed in vegetable oil (BSMO). *Fish. Aquatic Sci.* **2014,** *17,* 37–46.

Nanri, K.; Ogawa, J.; Nishikawa, T. Tunic of a pyurid ascidian *Microcosmus hartmeyeri* Oka is eaten locally in Japan. *The Nanki Seibutu* **1992,** *34,* 135.

Naranjo, S.A.; Kukurtçu, H.B.; Barbero, C.; Martin, S.; Carballo, J.L. Aquaculture of *Ecteinascidia turbinata* Herdman, 1880 as source of marine anticancer agents. In: Sawada, H., Yokosawa, H., Lambert, C.C. (eds). *The Biology of Ascidians*. Springer, Tokyo, **2001**. https://doi.org/10.1007/978-4-431-66982-1_54

Noguez, J.H.; Diyabalanage, T.K.K.; Miyata, Y.; Xie, X.S.; Valeriote, F.A.; Amsler, C.D.; McClintock, J.B.; Baker, B.J. Palmerolide macrolides from the Antarctic tunicate *Synoicum adareanum. Bioorg. Med. Chem.* **2011,** *19*, 6608–6614.

Numakunai, T.; Hoshino, Z. Biology of the ascidian, *Halocynthia roretzi* (Drasche), in Mutsu Bay. I. Differences of spawning time and external features. *Bull.Mar. Biol. Stn. Asamusi Tohoku Univ.* **1973,** *14,* 191–196.

Numakunai, T.; Hoshino, Z. Biology of the ascidian, *Halocynthia roretzi* (Drasche), in Mutsu Bay. II. One of three types which has the spawning season and the time different from two others. *Bull.Mar.Biol.Stn. Asamusi Tohoku Univ.* **1974,** *15,* 23–27.

Núñez-Pons, L.; Nieto, R.M.; Avila, C.; Jiménez, C.; Rodríguez, J. Mass spectrometry detection of minor new meridianins from the antarctic colonial ascidians *Aplidium falklandicum* and *Aplidium meridianum. J. Mass Spectrom.* **2015,** *50,* 103–111.

Oh, K.S.; Kim, J.S.; Heu, M.S. Food constituents of edible ascidians *Halocynthia roretzi* and *Pyura michaelseni. Korean J. Food Sci. Technol.* **1997,** *29,* 955–962.

Oh, Y.; Shim, K.; Ahn, C.; Kim, S.S.; Je, J. Sea squirt (*Halocynthia roretzi)* hydrolysates induce apoptosis in human colon cancer HT-29 cells through activation of reactive oxygen species. *Nutr. Cancer* **2019,** *71,* 118–127.

Oku, N.; Matsunaga, S.; Fusetani, N. Shishijimicins, A-C. Novel enediyne antitumor antibiotics from the ascidian *Didemnum proliferum. J. Am. Chem. Soc.* **2003**, *125,* 2044–2045.

Onder, F.C.; Ay, M. Chapter 9—Recent developments in natural bioactive peptides: Anticancer potential and structure–activity relationships. *Stud. Nat. Prod. Chem.* **2022,** *75,* 329–365.

Palanisamy, S.K.; Rajendran, N.M.; Marino, A. Natural products diversity of marine ascidians (Tunicates; Ascidiacea) and successful drugs in clinical development. *Nat. Prod. Bioprospect.* **2017***, 7,* 1–111.

Pan, Y.; Rodrigues, M.J.; Pereira, C.G.; Engrola, S.; Colen, R.; Mansinhos, I.; Romano, A.; Andrade, P.B.; Fernandes, F.; Custódio, L. Exploring the biotechnological value of marine invertebrates: a closer look at the biochemical and antioxidant properties of *Sabella spallanzanii* and *Microcosmus squamiger. Animals (Basel)* **2021**, *11,* 3557.

Papadopoulou, C.; Kanias, G.D. Tunicate species as marine pollution indicators. *Marine Poll. Bull.* **1977,** *8,* 229–231.

Park, C.K.; Matsui, T.; Watanabe, K.; Yamaguchi, K.; Konosu, S. Extractive nitrogenous constituents of two species of edible ascidians *Styela clava* and *S. plicata. Nippon Suisan Gakk* **1991,** *57,* 169–174.

Pavão, M.S. Unique sulfated polysaccharides from ascidians (Chordata, Tunicata). *Braz J Med. Biol. Res.* **1996,** *29,* 1227–1233.

Pavão, M.S.; Aiello, K.R.; Werneck, C.C.; Silva, L.C.; Valente, A.P.; Mulloy, B. et al. Highly sulfated dermatan sulfates from ascidians. Structure versus anticoagulant activity of these glycosaminoglycans. *J. Biol. Chem.* **1998,** *273,* 27848–27857.

Radhalakshmi, R.; Sivakumar, V.; Ali, H.A.F. Analysis of selected species of ascidians as bioindicators of metals in marine ecosystem. *Int. J. Curr. Microbiol. Appl. Sci.* **2014,** *3,* 755–764.

Rohmaha, Z.; Jeong, U.; Ticar, B.F.; Kimb, J.; Lee, J.; Kang, S.; Choi, B.D. Effect of ascidian (*Halocynthia roretzi*, Drasche 1884) tunics carotenoids on enhancing growth and muscle coloring of sea-reared rainbow trout (*Oncorhynchus mykiss,* Walbaum 1792) *Aquac. Rep.* **2016,** *4,* 22–29.

Roje-Busatto, R.; Ujevic, I. PSP toxins profile in ascidian *Microcosmus vulgaris* (Heller, 1877) after human poisoning in Croatia (Adriatic Sea). *Toxicon.* **2014***, 79,* 28–36.

Ryu, B.; Kim, S.K. Functional properties of ascidians by-products: Nutritional and medicinal value. In: Kim, SK. (eds.) *Seafood Processing By-Products.* Springer, New York, NY, **2014.**

Saha, A.; Yadav, R.; Rajendran, N. Biomaterials from sponges, ascidians and other marine organisms. *Int. J. Pharm. Sci. Rev. Res.* **2014,** *27,* 100–109.

Samuelsen, T.A.; Haustveit, G.; Kousoulaki, K. The use of tunicate (*Ciona intestinalis*) as a sustainable protein source in fish feed—Effects on the extrusion process, physical pellet quality and microstructure. *Anim. Feed Sci. Technol.* **2022,** 284, 1–13.

Santhanam, R.; Ramesh, S. *Biology and Ecology of Pharmaceutical Marine Tunicates*; CRC Press (Taylor & Francis), USA, **2019**, p. 181.

Sathishkumar, R.; Ananthan, G.; Raghunathan, C. Production and characterization of haloalkaline protease from ascidian-associated *Virgibacillus halodenitrificans* RSK CAS1 using marine wastes. *Ann. Microbiol.* **2015,** *65,* 1481–1493.

Seddiqi, H.; Oliaei, E.; Honarkar, H.; Jin, J.; Geonzon, L.C.; Bacabac, R.G.; Klein-Nulend, J. Cellulose and its derivatives: Towards biomedical applications. *Cellulose.* **2021,** *28,* 1893–1931.

Shenkar, N.; Swalla, B.J. Global diversity of ascidiacea. *PLoS One* **2011,** *6,* 1–12.

Suzuki, T.; Kubota, T.; Kobayashi, J. Eudistomidins H-K. New β-carboline alkaloids from the Okinawan marine tunicate *Eudistoma glaucus. Bioorg. Med. Chem. Lett.* **2011**, 21, 4220–4223.

Tamilselvi, M.; Sivakumar, V.; Ali, H.A.J.; Thilaga, R.D. Preparation of pickle from *Herdmania pallida*, simple ascidian. *World J. Dairy Food Sci.* **2010,** *5,* 88–92.

Tzafriri-Milo, R.; Benaltabet, T.; Torfstein, A.; Shenkar, N. The potential use of invasive ascidians for biomonitoring heavy metal pollution. *Front. Mar. Sci.* **2019,** *6,* doi:10.3389/fmars.2019.00611

Voogt, P.A.; van Rheenen, J.W. On the sterols of some ascidians. *Arch. Int. Physiol. Biochim.* **1975,** *83,* 563–572.

Watanabe, H.; Okawara, M.; Matahira, Y.; Mano, T.; Wada, T.; Suzuki, N.; Takara, T. The impact of ascidian *(Halocynthia roretzi)*-derived plasmalogen on cognitive function in healthy humans: A randomized, double-blind, placebo-controlled trial. *J. Oleo Sci.* **2020,** *69,* 1597–1607.

Watters, D.J. Ascidian toxins with potential for drug development. *Mar Drugs.* **2018,** *16,* 162.

Wawrzyniak, M.K.; Serrato, L.A.M.; Blanchoud, S. Artificial seawater based long-term culture of colonial ascidians. *Dev. Biol.* **2021a**, *480*, 91–104

Wawrzyniak, M.K.; Serrato, L.A.M.; Blanchoud, S. Long-term monitoring data logs of a recirculating artificial seawater based colonial ascidian aquaculture. *Data Brief.* **2021b,** *38,* 107372.

Yabu H, Ishibashi K, Grewal MS, Matsuo Y, Shoji N, Ito K. Bifunctional rare metal-free electrocatalysts synthesized entirely from biomass resources. Sci Technol Adv Mater. 2022 Jan 18;23(1):31-40. doi: 10.1080/14686996.2021.2020597.

Zhao, J.; Wei, J.; Liu, M.; Xiao, L.; Wu, N.; Liu, G.; Huang, H.; Zhang, Y.; Zheng, L.; Lin, X. Cloning, characterization and expression of a cDNA encoding a granulin-like polypeptide in *Ciona savignyi. Biochimie.* **2013**, *95,* 1611–1619.

Zhao, Y.; Li, J. Ascidian bioresources: Common and variant chemical compositions and exploitation strategy—Examples of *Halocynthia roretzi, Styela plicata, Ascidia* sp. and *Ciona intestinalis. Z. Naturforschung C.* **2016,** *71,* 165–180.

Zhu, Y.; Gao, H.; Han, S.; Li, J.; Wen, Q.; Dong, B. Antidiabetic activity and metabolite profiles of ascidian *Halocynthia roretzi. J. Funct. Foods* **2022,** *93,* 105095.

WEB REFERENCES

https://en.bdfish.org/2012/07/introduction-aquafeeds/ https://adoptandshop.org/the-invasive-nature-of-sea-squirts-understanding-their-biology- and-environmental-impacts/ https://australian.museum/learn/animals/sea-squirts/

https://blogs.ifas.ufl.edu/brevardco/2014/05/12/tunicates-aka-sea-squirts/

https://en.wikipedia.org/wiki/Bioremediation

https://en.wikipedia.org/wiki/Cynthiaol https://fit4art.com/en/blog/7051/Ascidians-nutritional-value-and-information
https://halkivisiting.proboards.com/thread/682/fouska-mock-ouster https://invert-embryo.blogspot.com/2013/06/ https://marinefeed.com/products-ciona-meal/
https://mybrainsupport.com/sea-pineapple/ https://neurosciencenews.com/aging-sea-squirts-20570/, 2022
https://notesfromdisgraceland.wordpress.com/hobbies/food
https://patents.google.com/patent/WO2013088177A1/en
https://pharmacy.ucsf.edu/news/2021/02/cancer-drug-stops-sars-cov-2-lab;
https://searchworks.stanford.edu/view/10773219
https://www.britannica.com/animal/tunicate
https://www.fiercebiotech.com/biotech/ascidian-therapeutics-rises-atps-ocean-floor-50-million-and-dynes-former-ceo
https://www.google.com/search?q=aphrodisiac+properties+of+Pyura+chilensis
https://www.notesonzoology.com/subphylum-urochordata/ascidia-external-features-and-metamorphosis/6271
https://www.osmosis.org/answers/trace-elements
https://www.robaid.com/bionics/marine-animal-tunicate-can-be-used-both-as-biofuel-and-fish-food.htm
https://www.sealifebase.se/summary/Pyura-chilensis.html
https://www.toppr.com/ask/question/ascidian-heart-is-unique-amongst-chordates-in-sense-that/
https://www.tus.ac.jp/en/mediarelations/archive/20220922_6725.html
https://www.umai-aomori.com/202106/sea-pineapple/
https://www.whoi.edu/science/B/people/kamaral/SeaSquirts.html
https://www.wipo.int/wipo_magazine/en/2009/01/article_0008.html
Bernard Picton . https://creativecommons.org/licenses/by/4.0/
Diego Delso. https://creativecommons.org/licenses/by-sa/4.0/
Jorge Stolfi -. https://creativecommons.org/licenses/by-sa/3.0/
Marcriusvil. https://creativecommons.org/licenses/by-sa/3.0/deed.en
Matthieu Sontag. https://creativecommons.org/licenses/by-sa/3.0/
Parent Géry. https://creativecommons.org/licenses/by-sa/3.0/deed.en
Rivas, https://www.osmosis.org/answers/trace-elements
Shriya https://www.notesonzoology.com/subphylum-urochordata/ascidia-external-features-and-metamorphosis/6271
Strohal et al. https://aslopubs.onlinelibrary.wiley.com/doi/pdf/10.4319/lo.1969.14.2.0265 tunicates-a-new-aquaculture-adventure).
Tatarinov,A.C. -. https://creativecommons.org/publicdomain/zero/1.0/deed.en
Univ of Guelph students. https://creativecommons.org/licenses/by/3.0/deed.en

INDEX

B

C

D

M

T